T0213988

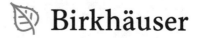

Compact Textbooks in Mathematics

This textbook series presents concise introductions to current topics in mathematics and mainly addresses advanced undergraduates and master students. The concept is to offer small books covering subject matter equivalent to 2- or 3-hour lectures or seminars which are also suitable for self-study. The books provide students and teachers with new perspectives and novel approaches. They may feature examples and exercises to illustrate key concepts and applications of the theoretical contents. The series also includes textbooks specifically speaking to the needs of students from other disciplines such as physics, computer science, engineering, life sciences, finance.

- **compact**: small books presenting the relevant knowledge
- **learning made easy:** examples and exercises illustrate the application of the contents
- **useful for lecturers:** each title can serve as basis and guideline for a semester course/lecture/seminar of 2–3 hours per week.

More information about this series at http://www.springer.com/series/11225

Marek Galewski

Basic Monotonicity Methods with Some Applications

 Birkhäuser

Marek Galewski
Institute of Mathematics
Lodz University of Technology
Łódź, Poland

ISSN 2296-4568   ISSN 2296-455X   (electronic)
Compact Textbooks in Mathematics
ISBN 978-3-030-75307-8   ISBN 978-3-030-75308-5   (eBook)
https://doi.org/10.1007/978-3-030-75308-5

Mathematics Subject Classification: 47H05

This book is published under the imprint Birkhäuser, www.birkhauser-science.com, by the registered company Springer Nature Switzerland AG.
The registered company address is: Gewerbestrasse 11, 6330 Cham, Switzerland

To my children, Katarzyna and Krzysztof,
who will not read beyond this page

Preface

The aim of this text is to introduce the reader to some basic tools from the theory of monotone operators together with some of their applications. Although our main concern is about the monotonicity approach, we also show some interplay between both the monotonicity and the variational approaches for which we provide a suitable background. The course material is basic and it does not assume any specific knowledge apart from some background in functional analysis (Banach and Hilbert spaces) and measure theory (mainly spaces of integrable functions) together with material from classic calculus courses. Such background since easily accessible elsewhere, for example in [5, 25, 55], is not included. Throughout the text we provide examples working only for ordinary differential equations while being aware that the world of partial equations, which is far more difficult to be explored, would provide the reader with some more inspiring applications. However, in order to keep the illustrating material relatively simple and not to concentrate on necessary space setting tools, we decided not to complicate some details. The course is meant for graduate students of mathematics who want to get some basics in the monotone operators theory and smoothly move to studying more advanced topics pertaining to more refined applications. Most of results are proved apart from some background theorems which we use as tools. A considerable number of exercises is included. These usually follow directly from methods discussed in this book and are sometimes necessary in order to understand what follows in the sequel. They serve also as examples, which is why the reader is advised to follow them at least by reading the content.

The material which is covered allows the reader to further study the topic of monotonicity theory by shifting to a more detailed exposition of pseudo-monotone operators and next to multivalued monotone operators. Presenting such topics with full details, as is aimed to be done throughout the text, would greatly extend our considerations.

These notes were used for courses at the Institute of Mathematics at Lodz University of Technology several times for some small groups of interested students. I would like to thank them for careful reading of the blackboard while a draft version of this course was presented. A lot of people contributed to the final version of this book. Dr Robert Stegliński pointed out some issues having read the final version.

Some improvements were also suggested by Dr Zdzisław Stempień, the retired assistant professor at my university. I would like to thank a lot my PhD student Michał Bełdziński for the careful reading of not only the final version but many preliminary notes as well.

Łódź, Poland   Marek Galewski
March 23, 2021

Contents

1   Introduction to the Topic of the Course 1
   1.1   Some Outline of the Problem Under Consideration 1
   1.2   The Finite Dimensional Monotonicity Methods 4
   1.3   Applications to Discrete Equations 11

2   Some Excerpts from Functional Analysis 15
   2.1   On the Weak Convergence 15
   2.2   On the Function Spaces 23
   2.3   On the du Bois-Reymond Lemma and the Regularity
       of Solutions 35
   2.4   Nemytskii Operator and the Krasnosel'skii Type Theorem 36
   2.5   Differentiation in Banach Spaces 39
   2.6   A Detour on a Direct Method in the Calculus of Variation 46

3   Monotone Operators 55
   3.1   Monotonicity 55
   3.2   On Some Properties of Monotone Operators 59
   3.3   Different Types of Continuity 65
   3.4   Coercivity 69
   3.5   An Example of a Monotone Mapping 71
   3.6   Condition (S) and Some Other Related Notions 75
   3.7   The Minty Lemma and the Fundamental Lemma for Monotone
       Operators 77

4   On the Fenchel-Young Conjugate 81
   4.1   Some Background from Convex Analysis 81
   4.2   On the Conjugate and Its Properties 85

5   Potential Operators 91
   5.1   Basic Concepts and Properties 91
   5.2   Invertible Potential Operators 98
   5.3   Criteria for Checking the Potentiality 101

6   Existence of Solutions to Abstract Equations 107
   6.1   Preliminary Result 107
   6.2   The Browder–Minty Theorem 109

6.3   Some Useful Corollaries 116
6.4   The Strongly Monotone Principle 117
6.5   Pseudomonotone Operators 118
6.6   The Leray–Lions Theorem 121

7   Normalized Duality Mapping 125
7.1   Introductory Notions and Properties 125
7.2   Examples of a Duality Mapping 130
       7.2.1   A Duality Mapping for $H_0^1 (0, 1)$ 130
       7.2.2   On a Duality Mapping for $L^p (0, 1)$ 131
       7.2.3   On a Duality Mapping for $W_0^{1,p} (0, 1)$ 132
7.3   On the Strongly Monotone Principle in Banach Spaces 132
7.4   On a Duality Mapping Relative to a Normalization Function 134

8   On the Galerkin Method 137
8.1   Basic Notions and Results 137
8.2   On the Galerkin and the Ritz Method for Potential Equations........ 140

9   Some Selected Applications 143
9.1   On Nonlinear Lax-Milgram Theorem and the Nonlinear
       Orthogonality 143
9.2   On a Certain Converse of the Lax-Milgram Theorem 147
9.3   Applications to the Differentiability of the Fenchel-Young
       Conjugate 148
9.4   Applications to Minimization Problems 150
9.5   Applications to the Semilinear Dirichlet Problem 154
       9.5.1   Examples and Special Cases 159
9.6   Applications to Problems with the Generalized p−Laplacian........ 160
9.7   Applications of the Leray–Lions Theorem 165
9.8   On Some Application of a Direct Method 169

References 175

Index 179

Introduction to the Topic of the Course

1

In this introductory chapter, we present some ideas concerning the solvability of nonlinear equations which we will develop further on. We start with a description of a Dirichlet problem which will be considered in the sequel once the proper functional setting has been given. For the time being, the problem that is sketched is located in the space of continuously differentiable functions. We will also introduce some preliminary abstract methodology now given in a finite dimensional setting—following partially [17] and [50]. Some simple application to the so-called algebraic equation finishes this section.

1.1 Some Outline of the Problem Under Consideration

As a simple example of what interests us in this book, let us consider the solvability of the following differential equation supplied with the so-called boundary conditions:

$$\begin{cases} -\ddot{u}(t) + a\dot{u}(t) = g(t) \text{ for } t \in (0, 1) \\ u(0) = u(1) = 0. \end{cases} \tag{1.1}$$

Here by $\dot{u}(t)$, we mean, for the time being, a classical derivative with respect to t. Such problem as (1.1) is called a **Dirichlet Problem** due to the requirement that function u becomes 0 at both ends of the interval. It differs from the initial value problem in the sense that the interval over which the solution exists is now fixed. In order to be able to determine which tools we need in the examination of the solvability of this problem, we must properly understand (1.1) and put it into a proper functional framework. When $g : [0, 1] \to \mathbb{R}$ is continuous and when a is some fixed real number, the natural space setting would be the following subspace

© The Author(s), under exclusive license to Springer Nature Switzerland AG 2021
M. Galewski, *Basic Monotonicity Methods with Some Applications*,
Compact Textbooks in Mathematics, https://doi.org/10.1007/978-3-030-75308-5_1

of $C^2 [0, 1]$ of twice continuously differentiable functions:

$$E = \left\{ u \in C^2 [0, 1] : u(0) = u(1) = 0 \right\}$$

equipped with a norm

$$\|u\|_{C^2} = \max_{t \in [0,1]} |u(t)| + \max_{t \in [0,1]} |\dot{u}(t)| + \max_{t \in [0,1]} |\ddot{u}(t)|.$$

Then E is a Banach space, but there is some drawback that it is not reflexive. The reflexivity is the property that we cannot get rid of in this book. Moreover, such a space framework would not allow for a discontinuous function g. This means that we need to seek for some other more suitable function space setting. We will undertake this task in detail in the sequel, but let us now have a look at how we can understand the meaning of a solution to a boundary value problem. Assume for the time being that some u is a solution to (1.1), and let us multiply (1.1) by a function $v \in C^1 [0, 1]$ also satisfying conditions $v(0) = v(1) = 0$. Next let us take integrals from 0 to 1 and perform integration by parts. Then making use of the boundary conditions that are imposed, we arrive at the following formula:

$$\int_0^1 \dot{u}(t) \dot{v}(t) \, dt - a \int_0^1 u(t) \dot{v}(t) \, dt = \int_0^1 g(t) v(t) \, dt. \tag{1.2}$$

It will appear soon that, for problems considered in this book, both (1.1) and (1.2) are equivalent in the sense that a function u (from a space that will be specified later on) solves (1.1) if and only if it solves (1.2). Thus our starting point could be formula (1.2), which is easier to be obtained. The idea of how to approach the solvability of (1.2) can be outlined as follows. Let

$$E_1 = \left\{ u \in C^1 [0, 1] : u(0) = u(1) = 0 \right\}.$$

Assume that u solves (1.2). Then we may define a linear and continuous functional on E_1 given by the following formula:

$$Av = \int_0^1 \dot{u}(t) \dot{v}(t) \, dt - a \int_0^1 u(t) \dot{v}(t) \, dt. \tag{1.3}$$

Later on we will use symbol $A(u)$ in order to underline that A depends on u, and also we will use symbol g^* in order to denote another linear and continuous functional on E_1

$$g^* v = \int_0^1 g(t) v(t) \, dt.$$

Then (1.2) may be written in a form of an abstract equation (by equating the functionals we have just defined):

$$A(u) = g^*. \tag{1.4}$$

Exercise 1.1 Prove that both (1.3) and (1.4) define continuous and linear functionals on E_1, which is equipped with the following norm:

$$\|u\|_{E_1} = \max_{t \in [0,1]} |u(t)| + \max_{t \in [0,1]} |\dot{u}(t)|. \tag{1.5}$$

Exercise 1.2 Prove that E_1 is a closed subspace of $C^1[0, 1]$ when considered with a norm (1.5).

This book is concerned with the solvability tools coined for such equations as given above. Here is some clue suggesting what will follow: if we sketch a graph of a real continuous monotone function $A : \mathbb{R} \to \mathbb{R}$ with infinite limits at both $+\infty$ and $-\infty$, then any line parallel to the x-axis intersects the graph and this intersection is a convex set (consisting of one point or being a closed interval). In this observation, the following properties are used:

(i) the continuity of A,
(ii) the monotonicity of A, and
(iii) the fact that $\lim_{|u| \to +\infty} |A(u)| = +\infty$.

Remembering these three main points in what follows, we will develop the theory leading to the existence of abstract equations and their further applications. We will address especially issues mentioned as (i) and (ii) via various types of monotonicity and continuity. In case A will appear invertible, then properties of its inversion will also be checked. Since certain types of equations like (1.4) may be obtained via equating to 0 the derivative of some functional, we will also include such consideration in our presentation. Condition (iii) is related to the so-called coercivity.

We must note at the end of this section that when both functions $f : [0, 1] \times \mathbb{R} \to \mathbb{R}$ and $g : [0, 1] \to \mathbb{R}$ are continuous, we may consider nonlinear versions of (1.1), like

$$\begin{cases} -\ddot{u}(t) + a\dot{u}(t) + f(t, u(t)) = g(t) \text{ for } t \in (0, 1) \\ u(0) = u(1) = 0, \end{cases}$$

and also, for some $p > 2$,

$$\begin{cases} -\frac{d}{dt}\left(\left|\frac{d}{dt}u(t)\right|^{p-2} \frac{d}{dt}u(t)\right) + f(t, u(t)) = g(t) \text{ for } t \in (0, 1) \\ u(0) = u(1) = 0. \end{cases}$$

Symbol $\frac{d}{dt}u(t)$ will be used to denote the derivative (usually understood in the a.e. sense) of a function u with respect to t. We need first to introduce some background properly prior to presenting more advanced problems.

1.2   The Finite Dimensional Monotonicity Methods

Let N be a fixed natural number. Symbol $(\cdot, \cdot)$ stands for the scalar product in $\mathbb{R}^N$ and $|\cdot|$ for the associated Euclidean norm. We will denote the scalar product in $\mathbb{R}^N$ as multiplication when it does not cause any confusion. We are concerned with the finite dimensional counterparts of (1.4) for which we describe how the monotonicity methods work. We shall start with equations considered on the real line and next pass to problems located in the N-dimensional space. Function $f : \mathbb{R} \to \mathbb{R}$ is called nondecreasing if for any $x, y \in \mathbb{R}$, it holds

$$x \leq y \Longrightarrow f(x) \leq f(y),$$

or equivalently, for any $x, y \in \mathbb{R}$, we have

$$(f(x) - f(y))(x - y) \geq 0. \tag{1.6}$$

In case

$$(f(x) - f(y))(x - y) > 0$$

for $x \neq y$, we say that f is increasing (strictly monotone).

Concerning the existence of solutions to equations in one variable, which involve monotone mappings and have just been mentioned above, we have the following results that stem from the intermediate value theorem:

Proposition 1.1

Assume that $f : (a, b) \to \mathbb{R}$ is continuous and nondecreasing, where $a \in \mathbb{R}$ or $a = -\infty$ and $b \in \mathbb{R}$ or $b = +\infty$. Then for any

$$y \in \left(\lim_{x \to a^+} f(x), \lim_{x \to b^-} f(x) \right), \tag{1.7}$$

equation

$$f(x) = y$$

has a solution that is located in some closed interval. The solution is unique in case f is increasing.

Exercise 1.3 Prove Proposition 1.1.

Proposition 1.2

Assume that $f : \mathbb{R} \to \mathbb{R}$ is continuous, nondecreasing, and

$$\lim_{x \to -\infty} f(x) = -\infty, \quad \lim_{x \to +\infty} f(x) = +\infty. \tag{1.8}$$

Then for any $y \in \mathbb{R}$, equation

$$f(x) = y$$

has a solution that is unique when f is increasing.

Exercise 1.4 Using Proposition 1.1, prove Proposition 1.2.

The assumption of monotonicity in Proposition 1.1 can be omitted if we assume that both limits in (1.7) exist. Condition (1.8) on the other hand pertains to the notion of coercivity, and it can be expressed as follows for mappings defined on $\mathbb{R}^N$:

Definition 1.1 (Coercive Map)

We say that $f : \mathbb{R}^N \to \mathbb{R}^N$ is coercive if

$$\lim_{|x| \to +\infty} \frac{(f(x), x)}{|x|} = +\infty.$$

Exercise 1.5 Prove that the mapping $f_1 : \mathbb{R}^2 \to \mathbb{R}^2$ given by

$$f_1(x_1, x_2) = (x_1, 2x_2)$$

is coercive, while $f_2 : \mathbb{R}^2 \to \mathbb{R}^2$ given by

$$f_2(x_1, x_2) = (2x_2, -x_1)$$

is not.

Exercise 1.6 Prove that the mappings given in the above exercise are norm coercive (or else weakly coercive), i.e.

$$\lim_{|x| \to +\infty} |f_i(x)| = +\infty, \text{ for } i = 1, 2.$$

The formula (1.6) provides a good starting point for a notion of a monotone operator which we now provide and which we supply with some exercises and properties.

Definition 1.2 (Monotone Map)

We say that $f : \mathbb{R}^N \to \mathbb{R}^N$ is monotone if for any $x, y \in \mathbb{R}^N$, we have

$$(f(x) - f(y), x - y) \geq 0.$$

Exercise 1.7 Consider the map $f : \mathbb{R}^2 \to \mathbb{R}^2$ defined by

$$f(x_1, x_2) = (x_2, -x_1).$$

Prove that f is monotone.

Exercise 1.8 Let A be a symmetric 2×2-matrix. Find conditions on A (in terms of both the entries and the eigenvalues) under which the mapping $f : \mathbb{R}^2 \to \mathbb{R}^2$ defined by $f(x) = Ax$ is monotone.

Proposition 1.3

Assume that $f : \mathbb{R}^N \to \mathbb{R}^N$ is surjective (i.e. $f(\mathbb{R}^N) = \mathbb{R}^N$) and monotone. Then f is weakly coercive, i.e.

$$\lim_{|x| \to +\infty} |f(x)| = +\infty. \tag{1.9}$$

Proof Supposing to the contrary, there is a sequence $(x_n)_{n=1}^{\infty} \subset \mathbb{R}^N$, $x_n \neq 0$, such that for some $M > 0$

$$|x_n| \to +\infty \text{ and } |f(x_n)| \leq M.$$

Define for $n \in \mathbb{N}$,

$$w_n = \frac{x_n}{|x_n|}.$$

Since sequence $(w_n)_{n=1}^{\infty}$ is bounded, we can take a convergent subsequence, which we denote by $(w_n)_{n=1}^{\infty}$. Let its limit be w_0. Since $f(\mathbb{R}^N) = \mathbb{R}^N$, we see that there is x_0 such that

$$f(x_0) = (M + 1) w_0.$$

Note that

$$\frac{x_n - x_0}{|x_n|} \to w_0 \text{ as } n \to +\infty \text{ and } \limsup_{n \to +\infty} |f(x_n)| \leq M.$$

The monotonicity of f implies that

$$\left(f(x_n), \frac{x_n - x_0}{|x_n|} \right) \geq \left(f(x_0), \frac{x_n - x_0}{|x_n|} \right).$$

Taking lim sup on both sides of the above inequality, we reach a contradiction.

Exercise 1.9 Show that function $f : \mathbb{R} \to \mathbb{R}$ defined by

$$f(x) = \begin{cases} x \sin x, & x \geq 0 \\ 0, & x < 0 \end{cases}$$

does not satisfy condition (1.9) but is surjective.

Our main goal in this section is to examine the solvability of the following nonlinear equation:

$$f(x) = a, \tag{1.10}$$

where $f : \mathbb{R}^N \to \mathbb{R}^N$ and where $a \in \mathbb{R}^N$ is fixed and next to relate the solvability of (1.10) to the monotonicity of f.

Lemma 1.1   (Finite Dimensional Minty Lemma)
Assume that $a \in \mathbb{R}^N$ is fixed and that a mapping $f : \mathbb{R}^N \to \mathbb{R}^N$ is continuous. If u solves the so-called variational inequality

$$(f(v) - a, v - u) \geq 0 \text{ for all } v \in \mathbb{R}^N, \tag{1.11}$$

then u also solves equation

$$f(x) = a. \tag{1.12}$$

If in addition f is monotone, then every solution to (1.12) is a solution to (1.11).

Proof Assume that u solves (1.11), and define $v = u + tw$ for $t > 0$ and any $w \in \mathbb{R}^N$. Then we have

$$(f(u + tw) - a, w) \geq 0 \text{ for all } w \in \mathbb{R}^N, \ t > 0.$$

Letting $t \to 0$, we obtain

$$(f(u) - a, w) \geq 0 \text{ for all } w \in \mathbb{R}^N.$$

Similarly putting $v = u - tw$ for $t > 0$, we see that also

$$(f(u) - a, w) \leq 0 \text{ for all } w \in \mathbb{R}^N.$$

Hence

$$(f(u) - a, w) = 0 \text{ for all } w \in \mathbb{R}^N,$$

and taking $w = f(u) - a$, we get the first assertion.

For the second part of the theorem, we observe that by the definition of the monotonicity and since (1.12) holds, we have

$$0 \leq (f(v) - f(u), v - u) = (f(v) - a, v - u)$$

for all $v \in \mathbb{R}^N$.

We have mentioned that solutions to equations involving monotone functions on the real line are located in some closed intervals or in other words closed and convex sets. Same assertion holds true for equations with monotone mappings in $\mathbb{R}^N$.

Lemma 1.2
Assume that $a \in \mathbb{R}^N$ is fixed and that $f : \mathbb{R}^N \to \mathbb{R}^N$ is continuous and monotone. The set K of solutions to (1.12) is closed and convex.

Proof For any fixed $v \in \mathbb{R}^N$, let us define a closed convex set

$$S_v = \left\{ u \in \mathbb{R}^N : (f(v) - a, v - u) \geq 0 \right\}. \tag{1.13}$$

Due to Lemma 1.1, we see that

$$K = \cap_{v \in \mathbb{R}^N} S_v.$$

Hence K is closed and convex as the intersection of closed and convex sets.

Exercise 1.10 Prove that set S_v defined by (1.13) is closed and convex.

Remark 1.1

Note that when we do not impose the coercivity assumption on f, we do not necessarily obtain that Eq. (1.12) is solvable as seen when $f(x) = \exp(x)$ and $a < 0$.

For the proof of the main existence result concerning equations involving monotone mappings, we need the Brouwer Fixed Point Theorem and one of its corollaries.

Theorem 1.1   (Brouwer Fixed Point Theorem)

Let $C \subset \mathbb{R}^N$ be closed, bounded, and convex. Suppose that $f : \mathbb{R}^N \to \mathbb{R}^N$ is continuous and that $f(C) \subseteq C$. Then f has a fixed point in C, i.e. there is $u \in C$ such that

$$f(u) = u.$$

Remark 1.2

The proof of the above fundamental theorem is not simple, and the methods involved are beyond the scope of this book. We refer, for example, to [18] for a proof via the topological degree. Some intuitive but involved approach is contained in [14]. The fixed point obtained in the Brouwer Theorem need not be unique.

Exercise 1.11   Prove the Brouwer Fixed Point Theorem when $N = 1$.

Lemma 1.3

Assume that $f : \mathbb{R}^N \to \mathbb{R}^N$ is a continuous mapping such that for some $R > 0$, it holds

$$(f(u), u) \geq 0 \text{ for } |u| = R. \tag{1.14}$$

Then there is $u_0 \in \mathbb{R}^N$ such that $f(u_0) = 0$ and $|u_0| \leq R$.

Proof Suppose that for all

$$u \in \overline{B(0, R)} := \left\{ a \in \mathbb{R}^N : |a| \leq R \right\},$$

we have $f(u) \neq 0$. Then we may define a continuous mapping $g : \overline{B(0, R)} \to \overline{B(0, R)}$ by

$$g(u) = -R \frac{f(u)}{|f(u)|}.$$

By the Brouwer Fixed Point Theorem, there is some $u_0 \in \overline{B(0, R)}$ such that $g(u_0) = u_0$. Then obviously $|u_0| = R$ and using (1.14), we obtain

$$0 < (u_0, u_0) = (g(u_0), u_0) = -R\left(\frac{f(u_0)}{|f(u_0)|}, u_0\right) \leq 0.$$

This contradiction finishes the proof.

Theorem 1.2   (Finite Dimensional Existence Theorem)
Assume that $a \in \mathbb{R}^N$ is fixed and that $f : \mathbb{R}^N \to \mathbb{R}^N$ is coercive, continuous, and monotone. Then the set K of solutions to (1.12) is non-empty, closed, and convex.

Proof Due to Lemma 1.2, we must only show that K is non-empty. We observe that we can assume $a = 0$ since mapping $g(\cdot) := f(\cdot) - a$ is continuous, coercive, and monotone as well. As for the coercivity, we note that from the Schwarz Inequality (i.e. $(x, y) \leq |x||y|$ for $x, y \in \mathbb{R}^N$), the following estimation holds for any $u \in \mathbb{R}^N$

$$\frac{(g(u), u)}{|u|} \geq \frac{(f(u), u)}{|u|} - |a|.$$

From the coercivity of g, there is some $R > 0$ such that for $|u| = R$, we have

$$(g(u), u) \geq 0.$$

Thus using Lemma 1.3, we obtain the assertion about the existence.

There is also a result about the existence and uniqueness which reads the following:

Theorem 1.3   (Finite Dimensional Existence and Uniqueness Theorem)
Assume that $a \in \mathbb{R}^N$ is fixed and that $f : \mathbb{R}^N \to \mathbb{R}^N$ is coercive, continuous, and strictly monotone. Then the set K of solutions to (1.12) is a singleton.

Proof The proof of the uniqueness follows by a contradiction and is left as an exercise.

The following exercise and example show clearly that one must take care when working with monotone operators even in a finite dimensional case.

Exercise 1.12 Check if there is an injective monotone function $f : \mathbb{R} \to \mathbb{R}$, which is not continuous and which is such that $f(\mathbb{R}) = \mathbb{R}$. Hint: recall that a monotone function of real variable has side limits at any point.

Example 1.1 Define $g : \mathbb{R} \to \mathbb{R}$ by

$$g(x) = \begin{cases} x, & x < 0, \\ x + 1, & x \geq 0. \end{cases}$$

Consider operator $f : \mathbb{R}^2 \to \mathbb{R}^2$ given by

$$f(x, y) = (y + g(x), -x).$$

Then f is monotone and surjective, but it is not continuous. The reader is invited to calculate all necessary details.

1.3   Applications to Discrete Equations

The theory of difference or in other words discrete equations is described in many excellent sources among which we suggest [33]. Concerning the application of the already introduced monotonicity tools, we investigate the so-called algebraic equation. This allows us to deal with discrete equations but at the same time to skip some details from the theory of difference equations that are not directly relevant to what we want to illustrate. For some information on such equations, the interested reader can refer to [2, 40].

Let us take a positive definite $N \times N$ real symmetric matrix

$$M = \begin{bmatrix} 2 & -1 & 0 & \dots & 0 \\ -1 & 2 & -1 & \dots & 0 \\ \dots & \dots & \dots & \dots & \dots \\ 0 & \dots & -1 & 2 & -1 \\ 0 & \dots & 0 & -1 & 2 \end{bmatrix}_{N \times N}$$

with eigenvalues $\lambda_1, \lambda_2, \dots, \lambda_N$ ordered as

$$0 < \lambda_1 \leq \lambda_2 \leq \cdots \leq \lambda_N$$

and consider a continuous function $f : \mathbb{R}^N \to \mathbb{R}^N$ such that

$$f(x_1, \ldots, x_N) = [f_1(x_1), \ldots, f_N(x_N)] \text{ for } x = (x_1, \ldots, x_N) \in \mathbb{R}^N,$$

where $f_k : \mathbb{R} \to \mathbb{R}$ for $k = 1, \ldots, N$.

We investigate the solvability of the following problem:

$$Mu = \lambda f(u) \tag{1.15}$$

subject to a positive parameter $\lambda > 0$ under the assumption that

A for each $k \in \{1, \ldots, N\}$, there exists a number $p_k > 0$ such that

$$(f_k(x) - f_k(y))(x - y) \geq p_k |x - y|^2$$

for $x, y \in \mathbb{R}$.

Problems like (1.15) are called algebraic equations or algebraic systems.

Theorem 1.4

*Let $p = \min_{1 \leq k \leq N} \{p_k\}$. If f satisfies condition **A**, then for each fixed $\lambda \in (\frac{\lambda_N}{p}, +\infty)$, problem (1.15) has a unique solution u^* in $\mathbb{R}^N$.*

Proof Define an operator $A : \mathbb{R}^N \to \mathbb{R}^N$ by

$$A(u) = \lambda f(u) - Mu.$$

Now Eq. (1.15) is equivalent to

$$A(u) = 0.$$

We see that obviously operator A is continuous. By assumption **A**, we get for $u, v \in \mathbb{R}^N$ that

$$(A(u) - A(v), u - v) = \lambda (f(u) - f(v), u - v) - (M(u - v), u - v) \geq (\lambda p - \lambda_N) |u - v|^2.$$

Thus A is a strictly monotone operator. The same argument shows that A is coercive. Therefore Theorem 1.3 implies that $A(u) = 0$ has a unique solution $u^* \in \mathbb{R}^N$, which proves our assertion.

Exercise 1.13 Let $p = \max_{1 \le k \le N}\{p_k\}$. Show that if for each $k \in \{1, \ldots, N\}$, there exists a number $p_k > 0$ such that

$$(f_k(x_1) - f_k(x_2))(x_1 - x_2) \le p_k |x_1 - x_2|^2$$

for all $x_1, x_2 \in \mathbb{R}$, then for each fixed $\lambda \in (0, \frac{\lambda_1}{p})$, problem (1.15) has unique solution in $\mathbb{R}^N$.

We need to mention that (1.15) serves as a discrete counterpart of (1.1) with $a = 0$. Such a correspondence is well described in [33] and leads to the investigation of the so-called non-spurious solutions to boundary value problems.

Some Excerpts from Functional Analysis

2

In this chapter we give some background on the function space setting which is based partially on [5, 25] and also [18]. We also provide some necessary information on differentiation in normed spaces, following [27] and the Nemytskii type operators after [16, 26]. Some background on the direct variational method and optimization are also given from [16, 29, 35]. Background from measure and integration may be recalled after [55] and some information on topics in real analysis in [46]. A thorough introduction not only to nonlinear functional analysis but also to all other topics covered in this textbook is to be found in [34, 42, 44], where also much more advanced topics are covered. In this chapter we provide concise but complete treatment necessary in what follows. However, we skip some details and do not go into too much detours. While we indicate several proofs, we give much more examples and also some explanations. We direct the interested Reader to the above-mentioned sources to study more deeply the material covered here, if necessary. Some versions of the text which appears in what follows although in a different form and with different approach towards the Sobolev spaces are to be found in the author's earlier book [21] (in Polish).

2.1 On the Weak Convergence

Let us fix some notation which will be used later on. Let E be a real Banach space. The space is separable if it contains a dense and countable subset. Let E^* denote the adjoint space, i.e. the space of all linear and continuous mappings defined on E with values in $\mathbb{R}$ which will be called functionals. The norm in E will be denoted by $\|\cdot\|$ and the norm in E^* by $\|\cdot\|_*$ so that to avoid any confusion. We would underline which norm we mean by writing a subscript when necessary. We will use notation $(\cdot, \cdot)_E$ for the scalar product in E in case it is a Hilbert space.

© The Author(s), under exclusive license to Springer Nature Switzerland AG 2021
M. Galewski, *Basic Monotonicity Methods with Some Applications*,
Compact Textbooks in Mathematics, https://doi.org/10.1007/978-3-030-75308-5_2

If $f \in E^*$, $x \in E$, then by $\langle f, x \rangle$ we mean a duality pairing between E^* and E, i.e. the action of a functional f on element x. Each element $x \in E$ defines also a linear and continuous functional on E^* via a duality pairing. If $E^{**} = (E^*)^*$ denotes the second dual, we can define a canonical embedding $\chi : E \to E^{**}$. In case such a map is surjective, we say that space E is reflexive. From now on we assume that E, if not said otherwise, is a real, reflexive, and separable Banach space.

We start with a definition of a weak convergence and some of its related properties.

Definition 2.1 (Weak Convergence)

A sequence $(x_n)_{n=1}^{\infty} \subset E$ is said to be weakly convergent to an element $x_0 \in E$ if for each $f \in E^*$ it holds

$$\langle f, x_n \rangle \to \langle f, x_0 \rangle \text{ as } n \to +\infty.$$

We will write

$$x_n \rightharpoonup x_0 \text{ in } E$$

and we will call element x_0 a weak limit of a sequence $(x_n)_{n=1}^{\infty}$.

The limit in the sense of a norm will be addressed as a strong limit. In this case we will write $x_n \to x_0$ in E which means

$$\lim_{n \to +\infty} \|x_n - x_0\| = 0.$$

The weak limit is uniquely defined. Indeed, if $x_n \rightharpoonup x_0$ and $x_n \rightharpoonup y_0$, then for any $f \in E^*$ we have

$$0 = \lim_{n \to +\infty} \langle f, x_n - x_n \rangle = \lim_{n \to +\infty} \langle f, x_n \rangle - \lim_{n \to +\infty} \langle f, x_n \rangle = \langle f, x_0 - y_0 \rangle.$$

This implies that $x_0 = y_0$. Moreover, a weakly convergent sequence is necessarily (norm) bounded, see Theorem 16.14 c) from [25].

Exercise 2.1 Show that in a finite dimensional space the weak and the strong convergence are equivalent.

Exercise 2.2 Show that if $x_n \to x_0$ in E, then $x_n \rightharpoonup x_0$ in E.

Exercise 2.3 Show that the weak limit obeys the same arithmetic laws as the strong one.

Remark 2.1

Since E is reflexive, the weak convergence in E^* is understood as follows: A sequence $(f_n)_{n=1}^{\infty} \subset E^*$ is said to be weakly convergent to an element $f_0 \in E^*$ if for each $x \in E$ it holds

$$\langle f_n, x \rangle \to \langle f_0, x \rangle \text{ as } n \to +\infty.$$

Definition 2.2

We say that a subset of E is sequentially weakly closed if it contains limits of its all weakly convergent sequences.

The above definition implies that any sequentially weakly closed set is (norm) closed. In space E an interval with ends $x_1, x_2 \in E$ is defined as follows:

$$[x_1, x_2] = \{x \in E : x = \alpha x_1 + (1 - \alpha)x_2, \ \alpha \in [0, 1]\}.$$

Set $C \subset E$ is called convex if each two distinct points in C can be connected by an interval contained in C. We are interested in determining when closed sets are also sequentially weakly closed. The answer is as follows (see Theorem 3.7 and also Corollary 3.8 in [5]):

Lemma 2.1

Let $D \subset E$ be closed and convex. Then D is sequentially weakly closed.

In order to make sure that a closed set need not be sequentially weakly closed let us consider the following example for which we need, however, some preparations. We recall firstly the Riesz Representation Theorem (for the proof the Reader may consult Theorem 5.5 from [5]):

Theorem 2.1 (Riesz Representation Theorem)

Let E be a real Hilbert space. There exists exactly one bijective mapping $R : E^* \to E$ such that:

(i) $(Rf, x)_E = \langle f, x \rangle$ for all $f \in E^*$ and $x \in E$;
(ii) $\|Rf\| = \|f\|_*$ for all $f \in E^*$.

Operator $R : E^* \to E$ is called the Riesz operator for space E. It allows one to introduce in E^* the scalar product $(f, g)_{E^*} = (Rf, Rg)_E$ thus making E^* into a Hilbert space, which can be identified with E by the Riesz operator R.

By l^2 we understand a space of sequences $(x_n)_{n=1}^{\infty} \subset \mathbb{R}$ such that

$$\sum_{n=1}^{\infty} x_n^2 < +\infty.$$

l^2 becomes a Hilbert space when it is endowed with the following scalar product:

$$(f, x)_{l^2} = \sum_{i=1}^{\infty} f_i x_i \text{ for } f, x \in l^2.$$

By the Riesz Representation Theorem it follows that $(l^2)^*$ is identified with l^2. Here is the announced example.

Example 2.1 Let us take a sequence from a unit sphere of l^2 given as follows:

$$e_1 = (1, 0, 0, \ldots), \ e_2 = (0, 1, 0, 0, ..), \ldots \tag{2.1}$$

This sequence is bounded and it does not contain any strongly convergent subsequence. On the other hand by a direct calculation using the necessary convergence condition from the series theory we demonstrate that (2.1) is weakly convergent to 0_{l^2}.

We can show that the convex hull of the set from the above example (i.e. the smallest convex set which contains it) becomes its weak closure. As for the weak compactness we have the following version of the Eberlein-Šchmulian Theorem (see Section 3.5 from [5] and Theorem 2.1.5 from [14] for the proof in a Hilbert space setting):

Theorem 2.2 (Sequential Weak Compactness of a Unit Ball)
The closed unit ball in E is sequentially weakly compact, i.e. each sequence from the ball contains a weakly convergent subsequence.

Further on we will need rather function than sequence spaces. We start with the simplest case. The space $L^1(0, 1)$ consists of Lebesgue integrable functions defined on $[0, 1]$ with values in $\mathbb{R}^N$ which when endowed with a norm

$$\|u\|_{L^1} = \int_0^1 |u(t)| \, dt$$

becomes a Banach space (although not reflexive). Similarly, $L^1([0, 1] \times [0, 1])$ consists of Lebesgue integrable functions defined on $[0, 1] \times [0, 1]$ with values in $\mathbb{R}^N$ endowed with a norm

$$\|u\|_{L^1} = \int\int_{[0,1]\times[0,1]} |u(t, s)| \, dt ds,$$

where symbol $\int\int_{[0,1]\times[0,1]}$ stands for the double integral. By $L^2(0, 1)$, denoted also L^2, we understand the space of all Lebesgue measurable functions defined on $[0, 1]$ with values in $\mathbb{R}^N$ which are square integrable. When endowed with a scalar product

$$(u, v)_{L^2} = \int_0^1 u(t) v(t) \, dt$$

for $u, v \in L^2(0, 1)$, it becomes a separable Hilbert space. A well known Schwarz Inequality (valid in any Hilbert space) is as follows:

$$(u, v)_{L^2} \le \|u\|_{L^2} \|v\|_{L^2}, \text{ for } u, v \in L^2(0, 1),$$

where

$$\|u\|_{L^2} = \sqrt{\int_0^1 |u(t)|^2 \, dt}.$$

Some properties of convergent (both strongly and weakly) sequences from $L^2(0, 1)$ now follow.

Remark 2.2

By the Riesz Representation Theorem we see that $\left(L^2(0, 1)\right)^*$ is identified with $L^2(0, 1)$. Take a sequence $(u_n)_{n=1}^\infty \subset L^2(0, 1)$ which is weakly convergent to some u_0, that is

$$\lim_{n \to +\infty} \int_0^1 u_n(t) v(t) \, dt = \int_0^1 u_0(t) v(t) \, dt \text{ for all } v \in L^2(0, 1).$$

Assume also that $(u_n)_{n=1}^\infty$ converges a.e. on $[0, 1]$ to some $\tilde{u}$, that is

$$\lim_{n \to +\infty} u_n(t) = \tilde{u}(t) \text{ for a.e. } t \in [0, 1].$$

Then $\tilde{u}(t) = u_0(t)$ for a.e. $t \in [0, 1]$. For the proof one may consult Lemma 1.19 from Chapter 2 in [19].

We will use the following property of weakly convergent sequences: Assume that $(x_n)_{n=1}^{\infty} \subset E$ and $(f_n)_{n=1}^{\infty} \subset E^*$. If either of these sequences converges weakly and the other one strongly to their respective limits $x_0 \in E$ and $f_0 \in E^*$, then it holds

$$\lim_{n \to +\infty} \langle f_n, x_n \rangle = \langle f_0, x_0 \rangle .$$

Note that the assumption that at least one of the above sequences converges strongly is crucial as well as the assumption that E is reflexive.

Exercise 2.4 Find examples of sequences $(x_n)_{n=1}^{\infty} \subset E$ and $(f_n)_{n=1}^{\infty} \subset E^*$ such that $x_n \rightharpoonup x_0$ and $f_n \rightharpoonup f_0$ and for which

$$\lim_{n \to +\infty} \langle f_n, x_n \rangle \neq \langle f_0, x_0 \rangle .$$

Some further important properties of space $L^2(0, 1)$ are also given below. In what follows $\chi_A : A \to \mathbb{R}$ denotes the indicator function of set $A \subset E$, i.e.

$$\chi_A(t) = \begin{cases} 1, & t \in A, \\ 0, & t \notin A. \end{cases}$$

Example 2.2 There exist non-constant sequences that are weakly and almost everywhere convergent at the same time. Define $(u_n)_{n=1}^{\infty} \subset L^2(0, 1)$ as follows:

$$u_n(t) = \sqrt{n} \chi_{\left(0, \frac{1}{n}\right)}.$$

Then $u_n \rightharpoonup 0$ and also $u_n(t) \to 0$ for a.e. $t \in [0, 1]$.

The above example can be further commented as follows:

Remark 2.3
If $(u_n)_{n=1}^{\infty} \subset L^2(0, 1)$ is norm convergent to some function $u_0 \in L^2(0, 1)$, then it contains a subsequence which is convergent a.e. on $[0, 1]$, again to the same function u_0. If $(u_n)_{n=1}^{\infty} \subset L^2(0, 1)$ is weakly convergent to some $u_0 \in L^2(0, 1)$, it does not mean that it contains any subsequence converging almost everywhere. Take, for example, sequence $(u_n)_{n=1}^{\infty} \subset L^2(0, 1)$ defined by $u_n(t) = \sin(2n\pi t)$ for $n \in \mathbb{N}$. From the Riemann–Lebesgue Lemma (recall the theory of Fourier expansions, for example, from [6]) we see that

$$\lim_{n \to +\infty} \int_0^1 \sin(2n\pi t) \, v(t) \, dt = 0 \text{ for any } v \in L^2(0, 1).$$

At the same time for any fixed $t \in (0, 1)$ sequence $(\sin(2n\pi t))_{n=1}^{\infty}$ obviously diverges.

Exercise 2.5 Prove the assertions in the above example and Remarks 2.2 and 2.3.

Exercise 2.6 Let E be a real Hilbert space. Let $(x_n)_{n=1}^{\infty} \subset E$ and let $x_0 \in E$. Show (using the properties of the scalar product) that if $x_n \rightharpoonup x_0$ and if $\|x_n\| \to \|x_0\|$, then $x_n \to x_0$.

We will need also some special types of Banach spaces which are defined as follows.

Definition 2.3 (Strictly Convex Space)

A normed linear space E is called strictly convex if the unit sphere contains no line segments on its surface, i.e. condition

$$\|x\| = 1, \ \|y\| = 1, \ x \neq y$$

implies that

$$\left\| \frac{1}{2} (x + y) \right\| < 1.$$

Definition 2.4 (Uniformly Convex Space)

A normed linear space E is called uniformly convex, if for each $\varepsilon \in (0, 2]$ there exists $\delta(\varepsilon) > 0$ such that if

$$\|x\| = 1, \|y\| = 1 \text{ and } \|x - y\| \geq \varepsilon,$$

then

$$\|x + y\| \leq 2(1 - \delta(\varepsilon)).$$

Exercise 2.7 Verify that $\mathbb{R}^2$ equipped with the Euclidean norm is uniformly convex, while it is not uniformly convex when we equip $\mathbb{R}^2$ with either

$$|x|_1 = |x_1| + |x_2| \tag{2.2}$$

or

$$|x|_2 = \max\{|x_1|, |x_2|\}. \tag{2.3}$$

Hint: draw unit balls in $\mathbb{R}^2$ with any of these norms.

Exercise 2.8 Show that $\mathbb{R}^2$ equipped with norms (2.2), (2.3) is not strictly convex as well.

Exercise 2.9 Prove that any real Hilbert space is uniformly convex. Hint: use the parallelogram law.

Exercise 2.10 Prove that a uniformly convex space is strictly convex.

Remark 2.4

A uniformly convex space is necessarily reflexive which follows by the Milman-Pettis Theorem, see Theorem 3.31 from [5]. The reflexivity of the space is often proved using its uniform convexity.

Now we proceed with some examples of uniformly convex spaces. In what follows, if not said otherwise, we take $p > 1$ and we put q such that

$$\frac{1}{p} + \frac{1}{q} = 1.$$

By $L^p(0, 1)$, denoted also L^p, we understand the space of all Lebesgue measurable functions defined on $[0, 1]$ with values in $\mathbb{R}^N$ which are integrable with power p. When endowed with a norm

$$\|u\|_{L^p} = \sqrt{\int_0^1 |u(t)|^p \, dt},$$

L^p becomes a uniformly convex separable Banach space. The convergence in L^p according to Remark 1.2.18 from [14] implies what follows: When $(u_n)_{n=1}^\infty \subset L^p$ is convergent to some $u_0 \in L^p$, then there is a subsequence $(u_{n_k})_{k=1}^\infty$ which converges a.e. on $[0, 1]$ and also a function $g \in L^p$ such that

$$|u_{n_k}(t)| \le g(t) \text{ for a.e. } t \in [0, 1], k \in \mathbb{N}.$$

The Riesz Representation Theorem in this case reads (see Theorem 4.11 in [5]):

Theorem 2.3

Let $1 < p < +\infty$ and let $\varphi \in (L^p(0, 1))^*$. Then there exists a unique function $u \in L^q(0, 1)$ such that

$$\langle \varphi, v \rangle_{(L^p)^*, L^p} = \int_0^1 u(t) v(t) \, dt \text{ for all } v \in L^p(0, 1)$$

and moreover

$$\|u\|_{L^p(0,1)} = \|\varphi\|_{(L^p(0,1))^*}.$$

From the above it follows that $(L^p (0, 1))^*$ is identified with $L^q (0, 1)$ and also

$$\langle u, v \rangle_{L^q, L^p} = \int_0^1 u(t) v(t) \, dt \text{ for all } u \in L^q (0, 1), \ v \in L^p (0, 1).$$

We will write rather $\langle u, v \rangle$ than $\langle u, v \rangle_{L^q, L^p}$ in the sequel. The Hölder Inequality holds for $u \in L^p$ and $v \in L^q (0, 1)$, that is

$$\int_0^1 |u(t) v(t)| dt \leq ||u||_{L^p} ||v||_{L^q}.$$

This implies that $uv \in L^1 (0, 1)$ and generalizes the Schwarz Inequality.

Another non-reflexive space is $L^\infty (0, 1)$. We say that a measurable function $u : [0, 1] \to \mathbb{R}^N$ belongs to $L^\infty (0, 1)$ if there exists a constant $c > 0$ such that

$$|u(t)| \leq c \text{ for a.e. } t \in [0, 1].$$

L^∞ becomes a Banach spaces when endowed with a following norm:

$$||u||_{L^\infty} = \inf \{c : |u(t)| \leq c \text{ for a.e. } t \in [0, 1]\}.$$

Exercise 2.11 Let $1 < p < q < +\infty$ and $u \in L^q (0, 1)$. Using the Hölder Inequality show that

$$||u||_{L^p} \leq ||u||_{L^q}.$$

2.2 On the Function Spaces

We are going here to put (1.1) into a proper functional setting. For this purpose we introduce the space of test functions

$$C_0^1 [0, 1] := \left\{ u \in C^1 [0, 1] : u(0) = u(1) = 0 \right\}.$$

From the theory of L^p spaces we know that $C_0^1 [0, 1]$ is dense in $L^p (0, 1)$ for all $1 \leq p < +\infty$. We define the Sobolev space $W^{1,p} (0, 1)$ as follows, see Chapter 8 in [5] for a very detailed treatment of topics discussed in this section. We will consider the case when $p > 1$.

Definition 2.5 (Weak Derivative)
We say that a function $u \in L^p (0, 1)$ belongs to $W^{1,p} (0, 1)$ if there is a function
$g \in L^p (0, 1)$ such that

$$\int_0^1 u(t) \dot{\varphi}(t) \, dt = - \int_0^1 g(t) \varphi(t) \, dt \tag{2.4}$$

for all $\varphi \in C_0^1 [0, 1]$. Function g is called a weak derivative of function u.

Exercise 2.12 Show that a weak derivative is uniquely defined.

If $u \in C^1 [0, 1]$, then (2.4) becomes a well known formula of integration by parts
and so g is a classical derivative of u. This is why we denote $\dot{u} = g$ and call it a weak
derivative or a derivative in a sense of the space $W^{1,p} (0, 1)$. The space $W^{1,p} (0, 1)$
is equipped with the following norm:

$$||u||_{W^{1,p}} = ||u||_{L^p} + ||\dot{u}||_{L^p} \tag{2.5}$$

or sometimes with an equivalent norm

$$||u||_{W^{1,p},2} = \left(||u||_{L^p}^p + ||\dot{u}||_{L^p}^p \right)^{1/p}. \tag{2.6}$$

Exercise 2.13 Prove that formulas (2.5), (2.6) define equivalent norms in $W^{1,p} (0, 1)$.

Example 2.3 A function

$$u(t) = \left| t - \frac{1}{2} \right|, \tag{2.7}$$

which is not differentiable in the classical calculus sense, belongs to $W^{1,p} (0, 1)$ for any
$p > 1$. We have

$$\dot{u}(t) = \begin{cases} -1, & 0 \le t < \frac{1}{2}, \\ 1, & \frac{1}{2} < t \le 1. \end{cases} \tag{2.8}$$

More generally, a continuous function on $[0, 1]$ that is piecewise C^1 on $[0, 1]$ belongs to
$W^{1,p} (0, 1)$ for any $p > 1$. Note on the other hand that $\dot{u}$ given above does not belong to
space $W^{1,p} (0, 1)$ for any $p > 1$ but it has a classical derivative for a.e. $t \in [0, 1]$. The only
candidate for a derivative in a sense of space $W^{1,p} (0, 1)$ is function $g \equiv 0$ for which formula
(2.4) does not hold for all test functions.

Exercise 2.14 Verify that the weak derivative of function (2.7) is given by (2.8). Show that
function given by (2.8) does not belong to $W^{1,p} (0, 1)$ for any $p > 1$.

When $p = 2$ we use the following notation:

$$H^1 (0, 1) := W^{1,2} (0, 1).$$

The space $H^1 (0, 1)$ is equipped with the scalar product

$$(u, v)_{H^1} = (u, v)_{L^2} + (\dot{u}, \dot{v})_{L^2} = \int_0^1 u(t) v(t) dt + \int_0^1 \dot{u}(t) \dot{v}(t) dt \text{ for } u, v \in H^1 (0, 1).$$

The corresponding norm which makes it into a Hilbert space is as follows:

$$\|u\|_{H^1} := \sqrt{\|u\|_{L^2}^2 + \|\dot{u}\|_{L^2}^2}. \tag{2.9}$$

Theorem 2.4

For any $p \in (1, 2) \cup (2, +\infty)$ the space $W^{1,p} (0, 1)$ is a reflexive and separable Banach space. The space $H^1 (0, 1)$ is a separable Hilbert space.

Proof We consider $W^{1,p} (0, 1)$ with a norm (2.5). Let $(u_n)_{n=1}^\infty \subset W^{1,p} (0, 1)$ be a Cauchy sequence. Then $(u_n)_{n=1}^\infty$ and $(\dot{u}_n)_{n=1}^\infty$ are Cauchy sequences in $L^p (0, 1)$. Moreover, $(u_n)_{n=1}^\infty$ converges in $L^p (0, 1)$ to some u_0, while $(\dot{u}_n)_{n=1}^\infty$ to some g. Then by formula (2.4) we have

$$\int_0^1 u_n(t) \dot{\varphi}(t) dt = -\int_0^1 \dot{u}_n(t) \varphi(t) dt \text{ for any } \varphi \in C_0^1 [0, 1].$$

Since both $\dot{\varphi}$ and φ belong obviously to $L^q (0, 1)$, we see taking the limits in the above that

$$\int_0^1 u_0(t) \dot{\varphi}(t) dt = -\int_0^1 g(t) \varphi(t) dt \text{ for any } \varphi \in C_0^1 [0, 1].$$

But this means that $g = \dot{u}_0$ and therefore $\|u_n - u_0\|_{W^{1,p}} \to 0$.

For the proof of reflexivity and separability we recall that $L^p (0, 1)$ is a reflexive and a separable Banach space. Therefore space

$$X = L^p (0, 1) \times L^p (0, 1)$$

equipped with a standard product norm

$$\|(u, v)\|_X = \|u\|_{L^p} + \|v\|_{L^p}$$

is reflexive and separable as well. Then operator

$$T : W^{1,p}(0, 1) \to X$$

defined by

$$T(u) = (u, \dot{u}) \text{ for } u \in W^{1,p}(0, 1)$$

is an isometry. Since $W^{1,p}(0, 1)$ is a Banach space, $T\left(W^{1,p}(0, 1)\right)$ is a closed subspace of X. Thus it is reflexive, see Theorem Proposition 3.20 from [5]. Similar arguments demonstrate that $W^{1,p}(0, 1)$ is separable.

Exercise 2.15 Prove that $H^1(0, 1)$ is a Hilbert space.

Now we turn to a result which says that elements of $W^{1,p}(0, 1)$ for any $p > 1$ are continuous (precisely speaking are equivalent to continuous functions). We need two auxiliary lemmas which we may address as regularity results (see also further Sect. 2.3 for some other point of view in this direction).

Lemma 2.2

Let $f \in L^1(0, 1)$ be such that

$$\int_0^1 f(t)\dot{\varphi}(t)\,dt = 0 \text{ for any } \varphi \in C_0^1[0, 1]. \qquad (2.10)$$

Then there is a constant $c \in \mathbb{R}^N$ such that $f(t) = c$ for a.e. $t \in [0, 1]$.

Proof We fix $\psi \in C[0, 1]$ such that $\int_0^1 \psi(t)\,dt = 1$. For any function $w \in C[0, 1]$ there exists $\varphi \in C_0^1[0, 1]$ such that

$$\dot{\varphi}(t) = w(t) - \left(\int_0^1 w(s)\,ds\right)\psi(t) \text{ for a.e. } t \in [0, 1].$$

Indeed, since

$$\int_0^1 \left(w(\tau) - \left(\int_0^1 w(s)\,ds\right)\psi(\tau)\right)d\tau = 0$$

we can define function $\varphi \in C_0^1[0, 1]$ as follows:

$$\varphi(t) = \int_0^t \left(w(\tau) - \left(\int_0^1 w(s)\,ds\right)\psi(\tau)\right)d\tau.$$

From (2.10) it follows that for any $w \in C[0, 1]$ we have

$$\int_0^1 f(t) \left(w(t) - \left(\int_0^1 w(s)\,ds \right) \psi(t) \right) dt = 0$$

which means that

$$\int_0^1 \left(f(t) - \left(\int_0^1 f(s) \psi(s)\,ds \right) \right) w(t)\,dt = 0.$$

The above implies that

$$f(t) - \left(\int_0^1 f(s) \psi(s)\,ds \right) = 0 \text{ for a.e. } t \in [0, 1].$$

Thus the assertion is proved.

In what follows we need a version the Fubini Theorem given so that to comply with our setting (the general version is e.g. in [55], Theorem 1.7.18):

Theorem 2.5 (Fubini Theorem)
Assume that $f \in L^1([0, 1] \times [0, 1])$. Then for a.e. $t \in [0, 1]$ function $s \mapsto f(t, s)$ belongs to $L^1(0, 1)$ and also function $t \mapsto \int_0^1 f(t, s)\,ds$ belongs to $L^1(0, 1)$. Similarly, for a.e. $s \in [0, 1]$ function $t \mapsto f(t, s)$ belongs to $L^1(0, 1)$ and also function $s \mapsto \int_0^1 f(t, s)\,dt$ belongs to $L^1(0, 1)$. Moreover, it holds

$$\iint_{[0,1] \times [0,1]} f(t, s)\,dt\,ds = \int_0^1 \left(\int_0^1 f(t, s)\,ds \right) dt = \int_0^1 \left(\int_0^1 f(t, s)\,dt \right) ds.$$

Using the above given Fubini Theorem we easily show that

Lemma 2.3
Let $f \in L^1(0, 1)$ and let $s \in [0, 1]$ be fixed. Then function $u : [0, 1] \to \mathbb{R}^N$ defined by the following formula:

$$u(t) = \int_s^t f(\tau)\,d\tau$$

is continuous and moreover

$$\int_0^1 u(t) \dot{\varphi}(t)\,dt = -\int_0^1 f(t) \varphi(t)\,dt \text{ for any } \varphi \in C_0^1[0, 1].$$

Theorem 2.6
Let $u \in W^{1,p}(0, 1)$. Then there is a function $u_0 \in C[0, 1]$ such that

$$u(t) = u_0(t) \text{ for a.e. } t \in [0, 1]$$

and moreover

$$u_0(t) - u_0(s) = \int_s^t \dot{u}(\tau) \, d\tau$$

for any $s, t \in [0, 1]$.

Proof We fix $s \in [0, 1]$ and define function $v : [0, 1] \to \mathbb{R}^N$ by the following formula:

$$v(t) = \int_s^t \dot{u}(\tau) \, d\tau.$$

Then from Lemma 2.3 it follows that

$$\int_0^1 v(t) \dot{\varphi}(t) \, dt = - \int_0^1 \dot{u}(t) \varphi(t) \, dt \text{ for any } \varphi \in C_0^1[0, 1].$$

Hence

$$\int_0^1 (v(t) - u(t)) \dot{\varphi}(t) \, dt = 0 \text{ for any } \varphi \in C_0^1[0, 1].$$

Thus by Lemma 2.2 we obtain the assertion.

Remark 2.5
The above theorem asserts that every function $u \in W^{1,p}(0, 1)$ admits one (and only one) absolutely continuous representative on $[0, 1]$ (see Remark 8, Chapter 8 in [5] for details). Thus we can replace u by its absolutely continuous representative which we will often do in what follows. We recall that an absolutely continuous function $u : [0, 1] \to \mathbb{R}^N$ has integrable a.e. derivative $\dot{u} : [0, 1] \to \mathbb{R}^N$ such that for all $t \in [0, 1]$ it holds

$$u(t) = u(0) + \int_0^t \dot{u}(s) \, ds.$$

(continued)

Remark 2.5 (continued)

Recall that any absolutely continuous function is continuous and for any absolutely continuous functions $f, g : [0, 1] \to \mathbb{R}^N$ we have integration by parts formula

$$\int_0^1 \dot{f}(t)g(t)dt = f(1)g(1) - f(0)g(0) - \int_0^1 f(t)\dot{g}(t)dt.$$

It follows that primitives of $L^p(0, 1)$ functions are in fact elements of $W^{1,p}(0, 1)$.

Remark 2.6

From the above remark it also follows that we can introduce $W^{1,p}(0, 1)$ as a space of such absolutely continuous functions whose derivatives (understood almost everywhere on $[0, 1]$) are integrable with power p. We will not apply this approach; however, it is also used in some background texts, compare with [35].

Note that since any $u \in W^{1,p}(0, 1)$ is absolutely continuous, the definition of the Sobolev space $W_0^{1,p}(0, 1)$ can be given as follows:

$$W_0^{1,p}(0, 1) := \{u \in W^{1,p}(0, 1), \ u(0) = 0, u(1) = 0\}.$$

The space $W_0^{1,p}(0, 1)$ is equipped with the following norm:

$$||u||_{W_0^{1,p}} = \left(\int_0^1 |\dot{u}(t)|^p \, dt \right)^{\frac{1}{p}}. \tag{2.11}$$

Exercise 2.16 Prove that formula (2.11) indeed defines a norm in $W_0^{1,p}(0, 1)$, while it does not define a norm in $W^{1,p}(0, 1)$.

Let us denote

$$||u||_C := \max_{t \in [0,1]} |u(t)|.$$

In case of the space $W_0^{1,p}(0, 1)$ we have the following Poincaré and Sobolev Inequalities.

Theorem 2.7
For any $u \in W_0^{1,p}(0,1)$ it holds

$$\|u\|_{L^p} \leq \|u\|_{W_0^{1,p}} \text{ and } \|u\|_C \leq \|u\|_{W_0^{1,p}}.$$

Moreover, the norms $\|\cdot\|_{W^{1,p}}$ and $\|\cdot\|_{W_0^{1,p}}$ are equivalent on $W_0^{1,p}(0,1)$.

Proof Let $u \in W_0^{1,p}(0,1)$. Then we have for any $t \in [0,1]$

$$|u(t)| = |u(t) - u(0)| = \left| \int_0^t \dot{u}(s)ds \right| \leq \|\dot{u}\|_{L^1}.$$

Applying the Hölder Inequality we see that

$$\|u\|_C \leq \|\dot{u}\|_{L^1} \leq \|\dot{u}\|_{L^p}. \tag{2.12}$$

We also have the obvious relation

$$\|u\|_{L^p} \leq \|u\|_C. \tag{2.13}$$

Combining inequalities (2.12) and (2.13) we obtain

$$\|u\|_{L^p} \leq \|\dot{u}\|_{L^p} = \|u\|_{W_0^{1,p}}.$$

Now, by the definition of the norm in $W^{1,p}(0,1)$ we see

$$\|u\|_{W_0^{1,p}} \leq \|u\|_{L^p} + \|\dot{u}\|_{L^p} = \|u\|_{W^{1,p}} \leq 2\|u\|_{W_0^{1,p}}.$$

This finishes the proof.

We proceed again to the case when $p = 2$. The space $H_0^1(0,1)$, denoted by H_0^1, is a closed subspace of $H^1(0,1)$ such that

$$H_0^1(0,1) := \left\{ u \in H^1(0,1) : u(0) = u(1) = 0 \right\}.$$

$H_0^1(0,1)$ is again a Hilbert space when equipped with a scalar product

$$(u,v)_{H_0^1} = \int_0^1 \dot{u}(t)\,\dot{v}(t)\,dt$$

and the associated norm

$$\|u\|_{H_0^1} = \sqrt{\int_0^1 |\dot{u}(t)|^2 \, dt}. \tag{2.14}$$

The Sobolev and the Poincaré inequalities read as follows: for any $u \in H_0^1(0,1)$ it holds

$$\|u\|_C \le \|u\|_{H_0^1}$$

and

$$\|u\|_{L^2} \le \frac{1}{\pi} \|u\|_{H_0^1}. \tag{2.15}$$

The Poincaré constant $\frac{1}{\pi}$ from (2.15) improves the constant from Theorem 2.7 for the case $p = 2$ and follows directly from the Parseval Identity known form the theory of Fourier expansions. For another derivation of this constant see Section 14.5 from [25].

From the Poincaré Inequality (2.15) it follows that on $H_0^1(0,1)$ the norm (2.9) is equivalent to (2.14). The space $H^2(0,1)$ consists of such elements of $H^1(0,1)$ whose derivatives are also elements of $H^1(0,1)$. In order to investigate the properties of weakly convergent sequences in $H_0^1(0,1)$ and $W_0^{1,p}(0,1)$ we need to recall the following compactness result, see Theorem 7.25 from [52].

Theorem 2.8 (Arzela–Ascoli Theorem)
Assume that the sequence of functions $f_n : [a,b] \to \mathbb{R}^N$ is uniformly bounded, i.e.

$$\exists_{M>0} \forall_{t \in [a,b]} \forall_{n \in \mathbb{N}} |f_n(t)| \le M$$

and uniformly equi-continuous, i.e.

$$\forall_{\varepsilon>0} \exists_{\delta>0} \forall_{n \in \mathbb{N}} \forall_{t,s \in [a,b]} |t - s| < \delta \implies |f_n(t) - f_n(s)| < \varepsilon.$$

Then sequence $(f_n)_{n=1}^{\infty}$ contains a subsequence which is uniformly convergent on $[a,b]$.

Theorem 2.9

Let $(u_n)_{n=1}^\infty \subset H_0^1 (0, 1)$ be weakly convergent to some $u_0 \in H_0^1 (0, 1)$. Then $u_n \rightrightarrows u_0$ on $[0, 1]$.

Proof Since $(u_n)_{n=1}^\infty \subset H_0^1 (0, 1)$ is weakly convergent, it is bounded, i.e. there is some constant $d > 0$ that $\|u_n\|_{H_0^1} \le d$. Fix $\varepsilon > 0$ and take $\delta < \left(\frac{\varepsilon}{d}\right)^2$. Then for

$$t, s \in [0, 1], \ s < t, \ |t - s| < \delta$$

we see that for any $n \in \mathbb{N}$

$$|u_n (t) - u_n (s)| \le \int_s^t |\dot{u}_n (\tau)|\, d\tau \le \sqrt{\int_s^t 1^2 d\tau} \sqrt{\int_s^t |\dot{u}_n (\tau)|^2\, d\tau}$$
$$= \sqrt{|t - s|} \sqrt{\int_0^1 |\dot{u}_n (\tau)|^2\, d\tau} < \varepsilon.$$

Then by the Arzela–Ascoli Theorem sequence $(u_n)_{n=1}^\infty$ has a subsequence $(u_{n_k})_{k=1}^\infty$ such that $u_{n_k} \rightrightarrows u_0$. Since any other uniformly convergent subsequence would also approach u_0, we see that the entire sequence $(u_n)_{n=1}^\infty$ is uniformly convergent.

Exercise 2.17 Prove that if $(u_n)_{n=1}^\infty \subset W_0^{1,p} (0, 1)$ is weakly convergent to $u_0 \in W_0^{1,p} (0, 1)$, then $u_n \rightrightarrows u_0$ on $[0, 1]$.

With the above preparation we may proceed to investigate space $W_0^{1,p} (0, 1)$ for $p > 1$ more deeply.

Exercise 2.18 Prove that $W_0^{1,p} (0, 1)$ is a closed subspace of $W^{1,p} (0, 1)$.

Theorem 2.10

For any $p \in (1, 2) \cup (2, +\infty)$ the space $W_0^{1,p} (0, 1)$ is a uniformly convex Banach space. $H_0^1 (0, 1)$ is a Hilbert space.

Proof We consider the case $p > 2$. For any $z, w \in \mathbb{R}^N$ the well known Clarkson Inequality holds (see Theorem 2.28 in [1])

$$\left|\frac{z + w}{2}\right|^p + \left|\frac{z - w}{2}\right|^p \le \frac{1}{2} \left(|z|^p + |w|^p\right).$$

Let $u, v \in W_0^{1,p}(0, 1)$ be such that

$$\|u\|_{W_0^{1,p}} = \|v\|_{W_0^{1,p}} = 1$$

and

$$\|u - v\|_{W_0^{1,p}} \geq \varepsilon \in (0, 2].$$

We have by the Clarkson Inequality

$$\left\|\tfrac{u+v}{2}\right\|_{W_0^{1,p}}^p + \left\|\tfrac{u-v}{2}\right\|_{W_0^{1,p}}^p = \int_0^1 \left(\left|\tfrac{\dot{u}(t)+\dot{v}(t)}{2}\right|^p + \left|\tfrac{\dot{u}(t)-\dot{v}(t)}{2}\right|^p\right) dt$$
$$\leq \tfrac{1}{2}\int_0^1 \left(|\dot{u}(t)|^p + |\dot{v}(t)|^p\right) dt = \tfrac{1}{2}\left(\|u\|_{W_0^{1,p}}^p + \|v\|_{W_0^{1,p}}^p\right) = 1.$$

Thus

$$\|u + v\|_{W_0^{1,p}} \leq 2\left(1 - \left(\tfrac{\varepsilon}{2}\right)^p\right)^{\frac{1}{p}}.$$

Therefore there exists

$$\delta(\varepsilon) = \left(1 - \left(1 - \left(\tfrac{\varepsilon}{2}\right)^p\right)^{\frac{1}{p}}\right) > 0$$

such that

$$\|u + v\|_{W_0^{1,p}} \leq 2(1 - \delta(\varepsilon)),$$

which proves the assertion.

Exercise 2.19 Prove that for $p \in (1, 2)$ the space $W_0^{1,p}(0, 1)$ is also uniformly convex. Hint: use the corresponding Clarkson Inequality:

$$\left\|\frac{u+v}{2}\right\|_{L^p}^q + \left\|\frac{u-v}{2}\right\|_{L^p}^q \leq \left(\frac{1}{2}\|u\|_{L^p}^p + \frac{1}{2}\|v\|_{L^p}^p\right)^{\frac{q}{p}}.$$

· Finally, we define space $H^2(0, 1)$ as follows:

$$H^2(0, 1) := \left\{u \in H^1(0, 1) : \dot{u} \in H^1(0, 1)\right\}.$$

Now we provide the form a dual space for both $H_0^1(0, 1)$ and $W_0^{1,p}(0, 1)$.

The dual space of $H_0^1(0, 1)$ is denoted by $H^{-1}(0, 1)$, H^{-1} for short. Using the Riesz Representation Theorem we identify L^2 with its dual directly. On the other hand $H_0^1 (0, 1)$ is identified by the Riesz Representation Theorem with $H^{-1} (0, 1)$ which is understood as follows. Let $F \in H^{-1}$, then there exists (not uniquely determined) $f \in L^2$ such that

$$\langle F, u \rangle_{H^{-1}, H_0^1} = \int_0^1 f(t) \dot{u}(t) \, dt$$

for all $u \in H_0^1 (0, 1)$ and also

$$\|F\|_{H^{-1}} = \|f\|_{L^2}.$$

We have the following continuous and dense inclusions:

$$H_0^1 (0, 1) \subset L^2 (0, 1) \subset H^{-1} (0, 1).$$

The dual space of $W_0^{1,p} (0, 1)$ is denoted by $W^{-1,q} (0, 1)$, $W^{-1,q}$ for short. Let $F \in W^{-1,q} (0, 1)$, then there exists (not uniquely determined) $f \in L^q$ such that

$$\langle F, u \rangle_{W^{-1,q}, W_0^{1,p}} = \int_0^1 f(t) \dot{u}(t) \, dt$$

for all $u \in W_0^{1,p}$ and

$$\|F\|_{W^{-1,q}} = \|f\|_{L^q}$$

Now we may return to the problem (1.1) and to its weak solvability, i.e. formula (1.2) under assumptions on the functions made in Sect. 1.1. Function $u \in H_0^1 (0, 1)$ will be called a weak solution to (1.1) if it satisfies

$$\int_0^1 \dot{u}(t) \dot{v}(t) \, dt - a \int_0^1 u(t) \dot{v}(t) \, dt = \int_0^1 g(t) v(t) \, dt$$

for all $v \in H_0^1 (0, 1)$. Equivalently we may write

$$\int_0^1 \dot{u}(t) \dot{v}(t) \, dt + a \int_0^1 \dot{u}(t) v(t) \, dt = \int_0^1 g(t) v(t) \, dt.$$

Now operator A from (1.4) is supposed to act from $H_0^1 (0, 1)$ into $H^{-1} (0, 1)$ which makes the setting more convenient. This means that to each fixed $u \in H_0^1 (0, 1)$ operator A assigns a linear and continuous functional, i.e. the element of $H^{-1} (0, 1)$. In order to check the relation between (1.4) and (1.1) we will pursue the question of the regularity of solutions.

2.3   On the du Bois-Reymond Lemma and the Regularity of Solutions

We proceed with a lemma which serves as a type of regularity result and which is called du Bois-Reymond Lemma or else the fundamental lemma of the calculus of variation for which we provide a somewhat different proof than contained in [35]. We have already established some related results in Lemmas 2.2 and 2.3 but it seems that what follows is of independents interest due to a very direct and simple proofs. We give some auxiliary lemma which follows directly from Lemma 2.2 and a main result which is supplied with some exercises.

> **Lemma 2.4   (Auxiliary Lemma)**
> *Let $h \in L^2(0, 1)$ be fixed and let*
>
> $$\int_0^1 h(t)\, \dot{v}(t)\, dt = 0 \text{ for all } v \in H_0^1(0, 1).$$
>
> *Then there exists $c \in \mathbb{R}^N$ such that $h(t) = c$ a.e. on $[0, 1]$.*

Exercise 2.20 Prove Lemma 2.4 directly. Hint: take $v(t) = \int_0^t (h(s) + c)\, ds$ for some c. Find c for which such a test function belongs to $H_0^1(0, 1)$.

> **Lemma 2.5   (du Bois-Reymond Lemma)**
> *Let functions $h \in L^2(0, 1)$, $f \in L^1(0, 1)$ be fixed and assume that*
>
> $$\int_0^1 (h(t)\, \dot{v}(t) + f(t)\, v(t))\, dt = 0 \text{ for all } v \in H_0^1(0, 1). \qquad (2.16)$$
>
> *Then function h is absolutely continuous and $\dot{h}(t) = f(t)$ a.e. on $[0, 1]$.*

Proof Note that $F : [0, 1] \to \mathbb{R}^N$ defined by

$$F(t) = \int_0^t f(s)\, ds$$

is absolutely continuous and that $\dot{F}(t) = f(t)$ a.e. on $[0, 1]$. Then we have

$$\int_0^1 \dot{F}(t)\, v(t)\, dt = -\int_0^1 F(t)\, \dot{v}(t)\, dt$$

and (2.16) has the following equivalent form:

$$\int_0^1 (h(t) - F(t))\, \dot{v}(t)\, dt = 0.$$

By Lemma 2.4 there exists a constant c such that $F(t) = h(t) + c$ for a.e. $t \in [0, 1]$. Now we see that the assertion follows.

Exercise 2.21 Show that if $f \in L^1(0, 1)$ is fixed and if

$$\int_0^1 f(t)\, v(t)\, dt = 0 \text{ for all } v \in H_0^1(0, 1),$$

then $f(t) = 0$ for a.e. $t \in [0, 1]$.

Exercise 2.22 Show that if $f \in C[0, 1]$ is fixed and if

$$\int_0^1 f(t)\, v(t)\, dt = 0 \text{ for all } v \in C_0^1[0, 1],$$

then $f(t) = 0$ for $t \in [0, 1]$.

Hint: argue by contradiction. Suppose that there is some $c \in (0, 1)$ that $f(c) > 0$. Consider the following test function:

$$v(t) = \begin{cases} (t - \alpha)(\beta - t), & t \in [\alpha, \beta], \\ 0, & t \notin [\alpha, \beta], \end{cases}$$

where $[\alpha, \beta] \subset (0, 1)$ is the interval on which f is positive.

Remark 2.7
Using Lemma 2.5 with $g \in L^2(0, 1)$ we see that any solution (1.2) is in fact in $H^2(0, 1) \cap H_0^1(0, 1)$, i.e. it solves (1.1) a.e. on $(0, 1)$. Moreover, when $g \in C[0, 1]$, the solution is twice continuously differentiable.

2.4 Nemytskii Operator and the Krasnosel'skii Type Theorem

The Krasnosel'skii type theorem will be used in proving the continuity for nonlinear operators and it concerns the so called Nemytskii operator.

Definition 2.6 (L^p-Carathéodory Function)

Let $p \geq 1$. We say that $f : [0, 1] \times \mathbb{R}^N \to \mathbb{R}^N$ is an L^p-Carathéodory function if the following conditions are satisfied:

(i) $t \mapsto f(t, x)$ is measurable on $[0, 1]$ for each fixed $x \in \mathbb{R}^N$,

(ii) $x \mapsto f(t, x)$ is continuous on $\mathbb{R}^N$ for a.e. $t \in [0, 1]$,

(iii) for each $d \in \mathbb{R}^+$ function $t \mapsto \max_{|x| \leq d} |f(t, x)|$ belongs to $L^p(0, 1)$.

When f satisfies only first two conditions of the above definition it is called a Carathéodory function. The following are examples of L^p-Carathéodory functions $f : [0, 1] \times \mathbb{R} \to \mathbb{R}$:

1. $f(t, x) = f_1(t) \cdot \arctan x$, $f_1 \in L^p(0, 1)$;
2. $f(t, x) = f_1(t)x^4 + f_2(t) x^3 + \arctan(t \cdot x)$, $f_1, f_2 \in f \in L^p(0, 1)$.

Let $\mathcal{M}$ be the set of all measurable functions $u : [0, 1] \to \mathbb{R}^N$. Composition of two measurable functions need not be measurable but when $u \in \mathcal{M}$ and $f : [0, 1] \times \mathbb{R}^N \to \mathbb{R}^N$ is a Carathéodory function, it follows that $t \mapsto f(t, u(t))$ is measurable. The Nemytskii operator induced by f is an operator

$$N_f : \mathcal{M} \to \mathcal{M}$$

such that

$$N_f(u)(\cdot) = f(\cdot, u(\cdot)) \text{ a.e. on } [0, 1].$$

We will need the Lebesgue Dominated Convergence Theorem for the proof of the next result.

Theorem 2.11 (Lebesgue Dominated Convergence Theorem)
Assume that $(f_n)_{n=1}^{\infty}$ is a sequence of measurable functions $f_n : [0, 1] \to \mathbb{R}^N$, for $n \in \mathbb{N}$, which is convergent a.e. on $[0, 1]$ to some function $f_0 : [0, 1] \to \mathbb{R}^N$. If there is a function $g \in L^p(0, 1)$ such that

$$|f_n(t)| \leq g(t) \text{ for a.e. } t \in [0, 1] \text{ and all } n \in \mathbb{N},$$

then $f_n, f_0 \in L^p(0, 1)$ for $n \in \mathbb{N}$ and also

$$\lim_{n \to +\infty} \|f_n - f_0\|_{L^p} = 0.$$

The following Theorem is rewritten following [26].

Theorem 2.12 (Generalized Krasnosel'skii Theorem)
Let $p_1, p_2 \geq 1$. Assume that $f : [0, 1] \times \mathbb{R}^N \to \mathbb{R}^N$ is a Carathéodory function. If for any sequence $(u_n)_{n=1}^\infty \subset L^{p_1}(0, 1)$ convergent to $\overline{u} \in L^{p_1}(0, 1)$ there exists a function $h \in L^{p_2}(0, 1)$ such that

$$|f(t, u_n)| \leq h(t), \text{ for } n \in \mathbb{N} \text{ and a.e. } t \in [0, 1], \tag{2.17}$$

then the Nemytskii operator induced by f

$$N_f : L^{p_1}(0, 1) \ni u(\cdot) \longmapsto f(\cdot, u(\cdot)) \in L^{p_2}(0, 1),$$

is well defined and continuous.

Proof Operator N_f is well defined by (2.17). We take any sequence $u_n \to \overline{u}$ in $L^{p_1}(0, 1)$. There exists a subsequence convergent almost everywhere, we denote it by the same symbol. Suppose that $(f(\cdot, u_n(\cdot)))_{n=1}^\infty$ does not converge to $f(\cdot, \overline{u}(\cdot))$ in $L^{p_2}(0, 1)$. Then there is $\varepsilon > 0$, such that for all N_0 there exist $n_0 \geq N_0$ that for all $n \geq n_0$

$$\int_0^1 |f(t, u_n(t)) - f(t, \overline{u}(t))|^{p_2} dt > \varepsilon. \tag{2.18}$$

Since f is a Carathéodory function we obtain that for *a.e.* $t \in [0, 1]$

$$\lim_{n \to +\infty} f(t, u_n(t)) = f(t, \overline{u}(t)).$$

By (2.17) it follows that

$$|f(t, \overline{u}(t))| \leq h(t) \text{ for a.e. } t \in [0, 1].$$

Using (2.17) and inequality

$$|a + b|^{p_2} \leq 2^{p_2-1}\left(|a|^{p_2} + |b|^{p_2}\right),$$

for $a, b \in \mathbb{R}^N$, we obtain

$$|f(t, u_n(t)) - f(t, \overline{u}(t))|^{p_2} \leq 2^{p_2} h(t)$$

for a.e. $t \in [0, 1]$. Hence by the Lebesgue Dominated Convergence Theorem we obtain

$$\lim_{n \to +\infty} \int_0^1 |f(t, u_n(t)) - f(t, \overline{u}(t))|^{p_2} dt = 0,$$

which contradicts (2.18).

There is also standard version of the above theorem which is usually called the Krasnosel'skii Theorem on the continuity of the Nemytskii operator:

Theorem 2.13 (Krasnosel'skii Theorem)
Let $p_1, p_2 \geq 1$. Let $f : [0, 1] \times \mathbb{R}^N \to \mathbb{R}^N$ be a Carathéodory function for which there exists a constant $b \geq 0$ and a nonnegative function $a \in L^{p_2}(0, 1)$ satisfying

$$|f(t, x)| \leq a(t) + b|x|^{\frac{p_1}{p_2}} \text{ for a.e. } t \in [0, 1] \text{ and all } x \in \mathbb{R}^N.$$

Then the Nemytskii operator $N_f : L^{p_1}(0, 1) \to L^{p_2}(0, 1)$ induced by f is well defined and continuous.

Remark 2.8
Note that the above theorem provides both necessary and sufficient condition for the continuity of the Nemytskii operator, see Theorem 2.4 in [16].

Exercise 2.23 Prove Theorem 2.13 directly and next using Theorem 2.12.

Remark 2.9
In Theorem 2.12 we could have assumed as well that N_f acts on a product of various L^p-spaces, L^∞ included. Hence when we refer to the generalized Krasnosel'skii Theorem we may also mean it in the sense of this remark.

2.5 Differentiation in Banach Spaces

Let X and Y be real Banach spaces. In what follows $\mathscr{L}(X, Y)$ stands for the space consisting of continuous linear mappings from X into Y. A mapping $f : X \to Y$ is said to be Gâteaux differentiable at $x_0 \in X$ if there exists $f'(x_0) \in \mathscr{L}(X, Y)$ such that for every $h \in X$

$$\lim_{t \to 0} \frac{f(x_0 + th) - f(x_0)}{t} = f'(x_0)h.$$

The operator $f'(x_0)$ is then called the Gâteaux derivative of f at x_0. A mapping f is continuously Gâteaux differentiable if

$$f' : X \ni x \mapsto f'(x) \in \mathscr{L}(X, Y)$$

is continuous in the relevant topologies.

Exercise 2.24 Show that the introduction of equivalent norms on X and Y does not influence the Gâteaux differentiability of $f : X \to Y$.

Function which is differentiable in the sense of Gâteaux need not be continuous as shows the following example:

Example 2.4 Let $f : \mathbb{R}^2 \to \mathbb{R}$ be given by

$$f(x, y) = \begin{cases} 1, & x = y^2, \ y > 0, \\ 0, & \text{otherwise}. \end{cases}$$

We see that f is Gâteaux differentiable at $(0, 0)$ with $f'(0, 0) = (0, 0)$ and it is obviously not continuous at $(0, 0)$.

Further on, we shall show that in some cases the Gâteaux differentiability of the functional implies that it is sequentially weakly lower semicontinuous. For Gâteaux derivatives one has the well known **Fermat Rule** which follows directly as in real line case.

> **Theorem 2.14 (Fermat Rule)**
> Let $F : X \to \mathbb{R}$. If $\overline{x} \in X$ is an argument of a minimum of F over X, i.e. if
>
> $$F(x) \geq F(\overline{x}) \ \text{for all } x \in X$$
>
> and if F is differentiable in the sense of Gâteaux (at least at $\overline{x}$), then for each $h \in X$ we have
>
> $$\left\langle F'(\overline{x}), h \right\rangle = 0.$$

An operator $f : X \to Y$ is said to be Fréchet differentiable at $x_0 \in X$ if there exists a continuous linear operator $f'(x_0) \in \mathscr{L}(X, Y)$ such that

$$\lim_{\|h\| \to 0} \frac{\|f(x_0 + h) - f(x_0) - f'(x_0)h\|}{\|h\|} = 0.$$

The operator $f'(x_0)$ is called the Fréchet derivative of f at x_0. When f is Fréchet differentiable it is also continuous and Gâteaux differentiable and obviously both derivatives coincide. A mapping f is continuously Fréchet differentiable if $f' : X \ni x \mapsto f'(x) \in \mathscr{L}(X, Y)$ is continuous. If f is continuously Gâteaux differentiable, then it is also continuously Fréchet differentiable. The space of all continuously Fréchet differentiable functions from X into Y will be denoted by $C^1(X, Y)$ or simply C^1 if there is no ambiguity about the spaces.

Remark 2.10

It easily follows from the definition, that $f' : X \to \mathscr{L}(X; Y)$ is continuous at $x_0 \in X$, if for any sequence $(x_n)_{n=1}^{\infty}$ such that $x_n \to x_0$ it holds

$$f'(x_n) h \to f'(x_0) h \text{ as } n \to +\infty$$

in Y uniformly with respect to h from the unit sphere in X, see Theorem 5.3 from [56]. This observation is used widely in proving that given functional is C^1 provided that it is differentiable in the sense of Gâteaux.

Exercise 2.25 Show that a function $f : \mathbb{R} \to \mathbb{R}$ defined by

$$f(t) = \begin{cases} t^2, & t \text{ rational,} \\ 0, & t \text{ irrational} \end{cases}$$

is Fréchet differentiable only at $t_0 = 0$.

We need to know how to differentiate the scalar product in the Hilbert space which is why we now assume that E is a Hilbert space in the Example that follows.

Example 2.5 We consider the following functional $F : E \to \mathbb{R}$ given by

$$F(x) = \frac{1}{2}(x, x)_E = \frac{1}{2}\|x\|^2.$$

Fix $x \in E$ and take any direction $h \in E$. We define $g : \mathbb{R} \to \mathbb{R}$ given by $g(t) = F(x + th)$ and observe that

$$g(t) = (x + th, x + th)_E = \frac{1}{2}\|x\|^2 + t(x, h)_E + \frac{1}{2}t^2\|h\|^2$$

which means that g is of class C^1 on $\mathbb{R}$ and therefore

$$g'(0) = (x, h)_E.$$

We can write the derivative as follows:

$$h \mapsto (x, h)_E .$$

Continuity of the derivative is obvious by the Riesz Representation Theorem.

We now provide an example and an exercise which clearly show that caution must be taken when differentiability of nonlinear maps is checked. Especially this is the case with the Nemytskii type operators.

Example 2.6 The mapping $f : L^2 (0, 1) \to L^2 (0, 1)$ defined by

$$f (u (\cdot)) = \sin (u (\cdot))$$

is not Fréchet differentiable, while it is everywhere Gâteaux differentiable. The arguments involved in proving this assertion require a more advanced study of the Nemytskii type operators than what we do here and thus we refer to [16] for details.

Exercise 2.26 Show that the mapping $f : H_0^1 (0, 1) \to L^2 (0, 1)$ defined by

$$f (u (\cdot)) = \sin (u (\cdot))$$

is continuously Fréchet differentiable and calculate the relevant derivative.

We differentiate composite mappings provided that the outer is at least Fréchet differentiable, see Theorem 3.2.12 from [14] for the proof.

Theorem 2.15 (Chain Rule)
Let X, Y, Z be real Banach spaces and assume that mapping $f : Y \to Z$ is differentiable in the sense of Fréchet while mapping $g : X \to Y$ is differentiable either in the sense of Fréchet or Gâteaux. Then mapping $f \circ g$ is differentiable at least in same the sense as g. Moreover, for any fixed $x \in X$ it holds that

$$(f \circ g)' (x) = f' (g (x)) g' (x) .$$

In case the outer mapping, i.e. f in the above formalism, is not differentiable in the sense of Fréchet, we may not apply the **Chain Rule** as seen by the following example.

Example 2.7 Let $X = Y = \mathbb{R}^2$, $Z = \mathbb{R}$ and let $\varphi : \mathbb{R}^2 \to \mathbb{R}^2$ be a C^1 mapping defined by

$$\varphi (x, y) = \left(x^2, y \right) .$$

Let $f : \mathbb{R}^2 \to \mathbb{R}$ be as in Example 2.4. Then mapping $g = f \circ \varphi$ has the following form:

$$g(x, y) = \begin{cases} 1, & |x| = y > 0, \\ 0, & \text{otherwise.} \end{cases}$$

We see that g in contrast to f is not differentiable in the sense of Gâteaux.

Example 2.8 With the **Chain Rule** we can differentiate the norm in a real Hilbert space E. We consider a functional $f : E \to \mathbb{R}$ given by

$$f(x) = \sqrt{(x, x)_E} = \|x\|.$$

By the Chain Rule for any $x \neq 0_E$ we see using Example 2.5 that

$$h \mapsto \frac{1}{\sqrt{(x, x)_E}} (x, h)_E$$

stands for the derivative of f at $x \neq 0_E$.

Finally, we introduce some second order sufficient convexity conditions which will be of use further on.

Definition 2.7 (n-th Gâteaux Variation)
Let $x_0 \in X$. Let $F : X \to \mathbb{R}$ be a functional and let $h \in X$ be fixed. Set $g : \mathbb{R} \to \mathbb{R}$ by

$$g(t) = F(x_0 + th).$$

Functional F is said to have the n-th Gâteaux variation at x_0 in direction h if the real function g is n-times differentiable at $t_0 = 0$. When F has n-th Gâteaux variation in any direction at x_0, we say that F has n-th Gâteaux variation at x_0. We denote the n-th Gâteaux variation in direction h at point x_0 by

$$F^{(n)}(x_0; h, \ldots, h).$$

We will still write $F'(x_0; h)$ instead of $F^{(1)}(x_0; h)$.

Example 2.9 When $L \in \mathscr{L}(X, Y)$, then it is C^1 and a derivative at any point equals L. Note that without continuity assumption a linear mapping admits only the first Gâteaux variation.

Now we proceed to some applications related to the convexity of a functional. Let $\alpha \in \mathbb{R}$ and let $F : X \to \mathbb{R}$. We define a set

$$F^\alpha = \{x \in X : F(x) \leq \alpha\}.$$

The epigraph of F is the following set:

$$\text{Epi}(F) = \{(x, \alpha) \in X \times \mathbb{R} : F(x) \leq \alpha\}.$$

Functional $F : X \to \mathbb{R}$ is convex on a convex subset $C \subset X$ if for any $x_1, x_2 \in C$, any $\lambda \in [0, 1]$ we have

$$F(\lambda x_1 + (1 - \lambda) x_2) \leq \lambda F(x_1) + (1 - \lambda) F(x_2). \tag{2.19}$$

In case $x_1 \neq x_2$, $\lambda \in (0, 1)$ and if it holds that

$$F(\lambda x_1 + (1 - \lambda) x_2) < \lambda F(x_1) + (1 - \lambda) F(x_2), \tag{2.20}$$

then F is called strictly convex.

Theorem 2.16 (Convexity Criteria)
Let $F : X \to \mathbb{R}$.

(i) *Then if F is convex it follows that for all $\alpha \in \mathbb{R}$ sets F^α are convex as well.*
(ii) *Then if $F : X \to \mathbb{R}$ is Gâteaux differentiable it follows that F is convex if and only if for all $x, y \in X$*

$$F(x) - F(y) \geq \left\langle F'(y), x - y \right\rangle. \tag{2.21}$$

Proof

(i) Noting that the empty set is convex by definition we may assume that F^α is non-empty. Take $x, y \in F^\alpha$ and $\lambda \in [0, 1]$. Then

$$F(\lambda x + (1 - \lambda) y) \leq \lambda F(x) + (1 - \lambda) F(y) \leq \lambda \alpha + (1 - \lambda) \alpha = \alpha.$$

Hence F^α is convex.

(ii) Let F be a Gâteaux differentiable convex functional and let $x, y \in X$, $\lambda \in (0, 1]$. We have

$$F((1 - \lambda)x + \lambda y) \leq (1 - \lambda) F(x) + \lambda F(y)$$

which implies that

$$F((1 - \lambda)x + \lambda y) \leq F(x) + \lambda(F(y) - F(x))$$

and

$$\frac{F((1-\lambda)x + \lambda y) - F(x)}{\lambda} \leq F(y) - F(x).$$

Now we easily calculate that

$$\langle F'(x), (y-x) \rangle = \lim_{\lambda \to 0^+} \frac{F((1-\lambda)x + \lambda y) - F(x)}{\lambda} \leq F(y) - F(x).$$

Now let us assume that (2.21) hold. Take $x, y \in X$ and $\lambda \in [0, 1]$. For pairs

$$(x, \lambda x + (1-\lambda)y), \ (y, \lambda x + (1-\lambda)y)$$

we have

$$F(x) - F(\lambda x + (1-\lambda)y) \geq \langle F'(\lambda x + (1-\lambda)y), (x - \lambda x - (1-\lambda)y) \rangle$$

and further

$$F(y) - F(\lambda x + (1-\lambda)y) \geq \langle F'(\lambda x + (1-\lambda)y), (y - \lambda x - (1-\lambda)y) \rangle.$$

This implies what follows

$$F(x) \geq F(\lambda x + (1-\lambda)y) + (1-\lambda) \langle F'(\lambda x + (1-\lambda)y), (x - y) \rangle \quad / \cdot \lambda$$

$$F(y) \geq F(\lambda x + (1-\lambda)y) + \lambda \langle F'(\lambda x + (1-\lambda)y), (y - x) \rangle \quad / \cdot (1-\lambda).$$

Summing up both sides we have

$$\lambda F(x) + (1-\lambda) F(y) \geq F(\lambda x + (1-\lambda)y).$$

Exercise 2.27 Check whether in part (ii) of the above theorem it suffice to use first Gâteaux variation instead of the Gâteaux derivative.

Remark 2.11

Note that convexity of F^α does not provide that F is convex as seen by the example of $F(x) = x^3$.

Theorem 2.17 (Sufficient Convexity Condition)
Assume that functional $F : X \to \mathbb{R}$ has a second variation on X. Then F if convex on X if and only if for any $u, h \in X$ it holds

$$F^{(2)} (u; h, h) \geq 0.$$

Moreover, if for any $u, h \in X$, $h \neq 0$, it holds

$$F^{(2)} (u; h, h) > 0,$$

then F is strictly convex.

For the proof of Theorem 2.17 we need the following version of the Mean Value Theorem (see Theorem 3.2.6 in [14]):

Theorem 2.18 (Mean Value Theorem)
Let $f : X \to Y$ have the first Gâteaux variation at all points of the $[a, b] \subset X$ in the direction of this segment, i.e. $f'(a + t(b - a); b - a)$ exists for all $t \in [0, 1]$. If the mapping $t \to f'(a + t(b - a); b - a)$ is continuous on $[0, 1]$, then

$$f(b) - f(a) = \int_0^1 f'(a + t(b - a); b - a)dt.$$

Exercise 2.28 Use the above given Mean Value Theorem in order to prove Theorem 2.17.

Exercise 2.29 Let E be a Hilbert space. Prove that functional $F : E \to \mathbb{R}$ defined by $F(x) = \frac{1}{2} \|x\|^2$ is strictly convex. Investigate the strict convexity of $F(x) = \|x\|^4$.

2.6   A Detour on a Direct Method in the Calculus of Variation

The variational approach towards the solvability of a nonlinear equation is based on the minimization of a certain action functional and linking the solution to equation considered with arguments of a minimum for a given functional. We proceed with some preliminary setting. Functional $J : E \to \mathbb{R}$ is lower semicontinuous on E if for all $x_0 \in E$ we have that

$$\liminf_{x \to x_0} J(x) \geq J(x_0);$$

where the above is understood as

$$\forall_{\varepsilon>0} \exists_{\delta>0} \forall_{x\in E} 0 < \|x - x_0\| < \delta \implies J(x) + \varepsilon \geq J(x_0).$$

The necessary and sufficient condition for the lower semicontinuity of J is that for each $\alpha \in \mathbb{R}$ set J^α is closed. Note that a sum of a continuous functional $J_1 : E \to \mathbb{R}$ and a lower semicontinuous $J_2 : E \to \mathbb{R}$ is lower semicontinuous since

$$\liminf_{x\to x_0} (J_1(x) + J_2(x)) = \lim_{x\to x_0} J_1(x) + \liminf_{x\to x_0} J_2(x).$$

The celebrated Weierstrass Theorem on the existence of an argument of a minimum of a continuous functional over a compact set reads as below:

Theorem 2.19 (Weierstrass Theorem)
Let $J : E \to \mathbb{R}$ be a lower semicontinuous functional and let $D \subset E$ be a compact set. Then J has at least one argument of a minimum over D.

The proof of the above result is precisely the same as in the standard calculus courses. The above-mentioned result has a drawback in infinite setting due to the fact that compact sets have empty interiors. Since both compactness and lower semicontinuity are basic assumptions used, we proceed as follows.

Definition 2.8 (Sequential Weak Lower Semicontinuity)
We say that a functional $J : E \to \mathbb{R}$ is sequentially weakly lower semicontinuous if for any $x_0 \in E$ and any sequence $(x_n)_{n=1}^\infty \subset E$ such that $x_n \rightharpoonup x_0$ it holds

$$\liminf_{n\to+\infty} J(x_n) \geq J(x_0).$$

Exercise 2.30 Prove that the equivalent condition to this provided in the above definition is to have all sets J^α sequentially weakly closed for $\alpha \in \mathbb{R}$.

Definition 2.9
We say that a functional $J : E \to \mathbb{R}$ is sequentially weakly continuous if for any $x_0 \in E$ and any sequence $(x_n)_{n=1}^\infty \subset E$ such that $x_n \rightharpoonup x_0$ it holds

$$\lim_{n\to+\infty} J(x_n) = J(x_0).$$

Exercise 2.31 Let $J : E \to \mathbb{R}$ be a sequentially weakly continuous functional. Prove that J is continuous.

Now we concentrate on the sufficient conditions for a functional to be sequentially weakly lower semicontinuous.

Theorem 2.20
Assume that functional $J : E \to \mathbb{R}$ is lower semicontinuous and convex. Then J is sequentially weakly lower semicontinuous.

Proof Since J is lower semicontinuous, it follows that sets J^α are closed for all $\alpha \in \mathbb{R}$. By Theorem 2.16 part (i) it follows that sets J^α are also convex for all $\alpha \in \mathbb{R}$. In view of Lemma 2.1 we see that since all J^α are closed and convex, these are sequentially weakly closed. But this implies that functional J is sequentially weakly lower semicontinuous.

Exercise 2.32 Prove that a norm in any Banach space is sequentially weakly lower semicontinuous.

Exercise 2.33 Prove that functional J is sequentially weakly lower semicontinuous if and only if set $\mathrm{Epi}(J)$ is sequentially weakly closed.

Theorem 2.20 has a direct application in proving that in uniformly convex spaces we have the following property:

Lemma 2.6
Let E be a uniformly convex space. Let $(x_n)_{n=1}^{\infty} \subset E$ and an element $x_0 \in E$. If $x_n \rightharpoonup x_0$ and $\|x_n\| \to \|x_0\|$, then $x_n \to x_0$.

Proof We can assume without loss of generality that $x_0 \neq 0$ and $x_n \neq 0$ for $n \in \mathbb{N}$. We set

$$y_0 = \frac{x_0}{\|x_0\|}, \ y_n = \frac{x_n}{\|x_n\|}.$$

Then $y_n \rightharpoonup y_0$ and $\|y_n\| \to \|y_0\| = 1$. Assume that $(y_n)_{n=1}^{\infty}$ does not converge in norm to y_0 and argue from this to a contradiction. Indeed, in this case for $\varepsilon \in (0, 2]$ we have a subsequence $(y_{n_k})_{k=1}^{\infty}$ such that $\|y_{n_k} - y_0\| \geq \varepsilon$. Since E is uniformly convex there exists $\delta(\varepsilon) > 0$ such that $\|y_{n_k} + y_0\| \leq 2(1 - \delta(\varepsilon))$. Letting $n_k \to +\infty$ and recalling that any norm as a convex and continuous functional is sequentially weakly lower semicontinuous, we see that $\|y_0\| \leq (1 - \delta(\varepsilon))$. But $\|y_0\| = 1$. The contradiction we have arrived at completes the proof of the lemma.

Now we proceed with some examples and counterexamples.

Example 2.10 Functional $J : l^2 \to \mathbb{R}$ defined by

$$J(x) = -\|x\|$$

is not sequentially weakly lower semicontinuous. If it were such, then $J_1(x) = \|x\|$ would be sequentially weakly upper semicontinuous and hence sequentially weakly continuous. This is not possible since for a sequence (2.1), which is weakly convergent to 0_{l^2}, we have that $J_1(e_n) = 1$.

Example 2.11 Let us consider $J : L^2(0, 1) \to \mathbb{R}$ given by

$$J(u) = \int_0^1 |u(t)|^2 \, dt.$$

Then according to the above example, J is not sequentially weakly continuous. However, as continuous and convex it is sequentially weakly lower semicontinuous. Considering the same functional on the space $H_0^1(0, 1)$, we see that now it is sequentially weakly continuous. Indeed, if $(u_n)_{n=1}^\infty$ is weakly convergent in $H_0^1(0, 1)$, then it follows that $u_n \to u_0$ in $L^2(0, 1)$. Now we obtain the assertion since J is continuous on $L^2(0, 1)$.

Let $\mathbb{R}_+ := [0, +\infty)$. The above example admits a more general version:

Example 2.12 Assume that $f : [0, 1] \times \mathbb{R}^N \to \mathbb{R}$ is a Carathéodory function and there exist $\phi \in L^\infty([0, 1], \mathbb{R}_+), h \in L^1(0, 1)$ such that

$$|f(t, x)| \le \phi(t) |x|^2 + h(t)$$

for a.e. $t \in [0, 1]$ and all $x \in \mathbb{R}^N$. Then functional $J : H_0^1(0, 1) \to \mathbb{R}$ defined by

$$J(u) = \int_0^1 f(t, u(t)) \, dt$$

is sequentially weakly continuous. A sequence $(u_n)_{n=1}^\infty$ which is weakly convergent in $H_0^1(0, 1)$ to some u_0 converges also uniformly on $[0, 1]$. Thus there is some $d > 0$ that $\|u_n\|_C \le d$. This further implies that

$$|f(t, u_n(t))| \le d^2 |\phi(t)| + |h(t)|$$

for a.e. $t \in [0, 1]$. Hence applying the Lebesgue Dominated Convergence Theorem we obtain the assertion.

The following exercises are counterparts of the above-mentioned examples in the case of space $W_0^{1,p}(0, 1)$.

Exercise 2.34   Consider $J : W_0^{1,p}(0, 1) \to \mathbb{R}$ given by

$$J(u) = \int_0^1 |u(t)|^p \, dt.$$

Prove that f is sequentially weakly continuous. Consider the same functional on $L^p(0, 1)$ and comment on its continuity and sequential weak continuity.

Exercise 2.35   Assume that $f : [0, 1] \times \mathbb{R}^N \to \mathbb{R}$ is a Carathéodory function and there exist $\phi \in L^\infty(0, 1), h \in L^1(0, 1)$ such that

$$|f(t, x)| \le \phi(t) |x|^p + h(t)$$

for a.e. $t \in [0, 1]$ and all $x \in \mathbb{R}^N$. Prove that functional $J : W_0^{1,p}(0, 1) \to \mathbb{R}$ defined by

$$J(u) = \int_0^1 f(t, u(t)) \, dt$$

is sequentially weakly continuous.

Now, having described the background, we provide the following counterpart of the Weierstrass Theorem:

Theorem 2.21
Let $J : E \to \mathbb{R}$ be sequentially weakly lower semicontinuous and let $D \subset E$ be sequentially weakly compact. Then J has at least one argument of a minimum over D.

Proof Suppose J is not bounded from below on D. Then there is a sequence $(x_n)_{n=1}^\infty \subset D$ such that $\lim_{n \to +\infty} J(x_n) = -\infty$. Observe that $(x_n)_{n=1}^\infty$ has a weakly convergent subsequence $(x_{n_k})_{k=1}^\infty$ with a weak limit $\tilde{x}$. Hence $\lim_{k \to +\infty} J(x_{n_k}) = -\infty$ and since f is sequentially weakly lower semicontinuous we see that

$$-\infty = \lim_{k \to +\infty} J(x_{n_k}) \ge J(\tilde{x}),$$

which is impossible. Hence J is bounded from below on D and it has a weakly convergent minimizing sequence $(x_n)_{n=1}^\infty \subset D$ with a weak limit x_0. We then have

$$\inf_{x \in D} J(x) = \liminf_{n \to +\infty} J(x_n) \ge J(x_0) \ge \inf_{x \in D} J(x).$$

Then x_0 is the argument of a minimum of functional J over D.

In case $D = E$, the boundedness of any minimizing sequence is guaranteed when sets J^α for each $\alpha \in \mathbb{R}$ are bounded. Such sets are bounded when functional J is (weakly) coercive, i.e. when

$$\lim_{\|x\| \to +\infty} J(x) = +\infty.$$

This leads to the following version of the Weierstrass Theorem which is used to derive the so-called direct method of the calculus of variations.

Theorem 2.22   (Direct Method of the Calculus of Variations)
Let functional $J : E \to \mathbb{R}$ be Gâteaux differentiable, sequentially weakly lower semicontinuous and coercive. Then J has at least one argument of a minimum x_0 over E, $J(x_0) = \inf_{x \in E} J(x)$, which is also a critical point, that is

$$\langle J'(x_0), h \rangle = 0 \text{ for all } h \in E. \tag{2.22}$$

Proof Take such α that J^α is non-empty. Since functional J is coercive it follows that J^α is bounded. We see at once that

$$\inf_{x \in E} J(x) = \inf_{x \in J^\alpha} J(x).$$

Using $D = J^\alpha$ we can apply Theorem 2.21 in order to find an element $x_0 \in J^\alpha$ such that $\inf_{x \in J^\alpha} J(x) = J(x_0)$. Hence x_0 is an argument of a minimum of J over E. When we apply the **Fermat Rule** the proof is finished.

Equation (2.22) is often called the Euler–Lagrange equation for functional J. We conclude this section with some easy application of Theorem 2.21 to functional

$$J : H_0^1(0, 1) \to \mathbb{R}$$

given by

$$J(u) = \frac{1}{2} \int_0^1 |\dot{u}(t)|^2 \, dt + \frac{1}{4} \int_0^1 |u(t)|^4 \, dt + \int_0^1 g(t) u(t) \, dt, \tag{2.23}$$

where $g \in L^2(0, 1)$ is fixed. We see at once that J is well defined, i.e. for any fixed $u \in H_0^1(0, 1)$ we have that $J(u)$ is finite.

Exercise 2.36 Show that functional J given by (2.23) is continuously differentiable and (strictly) convex.

Theorem 2.23
Let $g \in L^2(0, 1)$. Then functional J given by (2.23) has exactly one critical point which is an argument of a minimum.

Proof Since J is convex and continuous by Theorem 2.20 it follows that J is sequentially weakly lower semicontinuous over $H_0^1(0, 1)$. Since J is strictly convex, it has at most one argument of a minimum. Thus in order to apply Theorem 2.22 it suffices to show that J is coercive.

From the Schwarz Inequality and the Poincaré Inequality we see that

$$\left| \int_0^1 g(t) u(t)\, dt \right| \leq \|u\|_{L^2} \|g\|_{L^2} \leq \frac{1}{\pi} \|u\|_{H_0^1} \|g\|_{L^2} \text{ for all } u \in H_0^1(0, 1)$$

and next also

$$\frac{1}{4} \int_0^1 u^4(t)\, dt + \int_0^1 u(t) g(t)\, dt \geq -\frac{1}{\pi} \|u\|_{H_0^1} \|g\|_{L^2} \text{ for all } u \in H_0^1(0, 1).$$

Thus it holds that

$$J(u) \geq \frac{1}{2} \|u\|_{H_0^1}^2 - \frac{1}{\pi} \|u\|_{H_0^1} \|g\|_{L^2} \text{ for all } u \in H_0^1(0, 1).$$

Hence

$$J(u) \to +\infty \text{ as } \|u\|_{H_0^1} \to +\infty.$$

Therefore by Theorem 2.22 there exists exactly one $u_0 \in H_0^1(0, 1)$ such that

$$\left\langle J'(u_0), h \right\rangle = 0 \text{ for all } h \in H_0^1(0, 1). \tag{2.24}$$

Remark 2.12
Relation (2.24) means that

$$\int_0^1 \dot{u}_0(t) \dot{h}(t)\, dt + \int_0^1 \left(u_0^3(t) + g(t) \right) h(t)\, dt = 0 \text{ for all } h \in H_0^1(0, 1).$$

(continued)

Remark 2.12 (continued)

Using the du Bois-Reymond Lemma we see from the above that function $\dot{u}_0$ has a derivative for a.e. $t \in [0, 1]$ which is integrable with square and it holds

$$\ddot{u}_0(t) = u_0^3(t) + g(t) \text{ for a.e. } t \in [0, 1].$$

Hence $u_0 \in H_0^1(0, 1) \cap H^2(0, 1)$ is the unique solution of the following differential equation:

$$\ddot{u}(t) = u^3(t) + g(t)$$

with boundary conditions

$$u(0) = u(1) = 0.$$

Exercise 2.37 Follow the lines of the above example if $g \in L^1(0, 1)$ and if $g \in C(0, 1)$. What can we say about the solution in this case?

Exercise 2.38 Let $G : \mathbb{R} \to \mathbb{R}$ be a continuously differentiable convex function with a derivative $g : \mathbb{R} \to \mathbb{R}$ and consider a functional

$$J : H_0^1(0, 1) \to \mathbb{R}$$

given by

$$J(u) = \frac{1}{2} \int_0^1 |\dot{u}(t)|^2 \, dt + \int_0^1 G(u(t)) \, dt. \tag{2.25}$$

Prove that J is differentiable in the sense of Gâteaux, sequentially weakly lower semicontinuous and coercive. Argue that J has exactly one minimizer. Hint: while proving the coercivity show that the following estimation holds

$$G(x) \geq -|g(0)| \, |x| + G(0)$$

for all $x \in \mathbb{R}$.

Exercise 2.39 Find the Dirichlet problem which is satisfied by the argument of a minimum to functional J given by (2.25).

Monotone Operators

3

In this chapter, we base on [19] and [17] adding some examples and ideas from [14] and [10, 11, 37]. The monotonicity methods have been started in [30, 31] and also in [38, 39]. The historical remarks concerning monotone operators are to be found in [3]. Some notions introduced in the finite dimensional setting are next repeated in the general one just for the completeness of the presentation.

3.1 Monotonicity

Recall that, if not said otherwise, E is a real, separable, and reflexive Banach space.

Definition 3.1 (Different Types of Monotonicity)
Operator $A : E \to E^*$ is called

  (i) monotone, if for all $u, v \in E$, it holds

$$\langle A(u) - A(v), u - v \rangle \geq 0;$$

  (ii) strictly monotone, if for all $u, v \in E, u \neq v$, it holds

$$\langle A(u) - A(v), u - v \rangle > 0;$$

  (iii) uniformly monotone, if there exists an increasing function $\rho : [0, +\infty) \to [0, +\infty)$ such that $\rho(0) = 0$ and for all $u, v \in E$

$$\langle A(u) - A(v), u - v \rangle \geq \rho(\|u - v\|) \|u - v\|;$$

© The Author(s), under exclusive license to Springer Nature Switzerland AG 2021
M. Galewski, *Basic Monotonicity Methods with Some Applications*,
Compact Textbooks in Mathematics, https://doi.org/10.1007/978-3-030-75308-5_3

(iv) strongly monotone (called also m-strongly monotone), if there exists a constant $m > 0$ such that for all $u, v \in E$, it holds

$$\langle A(u) - A(v), u - v \rangle \geq m \, \|u - v\|^2 \, ;$$

(v) d-monotone, if for some increasing function $\rho : [0, +\infty) \to \mathbb{R}$, it holds for all $u, v \in E$

$$\langle A(u) - A(v), u - v \rangle \geq (\rho(\|u\|) - \rho(\|v\|)) (\|u\| - \|v\|). \tag{3.1}$$

A monotone operator sends points from E into functionals from E^*. Continuity of an operator A is understood as usual, i.e. $u_n \to u_0$ in E implies that $A(u_n) \to A(u_0)$ in E^*. A few simple examples of monotone operators as well as exercises now follow in order to acquaint the reader with this notion.

Example 3.1 Let E be Hilbert space. Let $A : E \to E$ be linear and such that

$$(A(u), u)_E \geq 0 \text{ for } u \in E.$$

Then A is monotone. Indeed, for any $u, v \in E$ we have

$$(A(u) - A(v), u - v)_E = (A(u - v), u - v)_E \geq 0.$$

Example 3.2 Take the operator $A : H_0^1(0, 1) \to H^{-1}(0, 1)$ defined by

$$\langle A(u), v \rangle = \int_0^1 \dot{u}(t) \, \dot{v}(t) \, dt \text{ for } u, v \in H_0^1(0, 1).$$

Using the Schwarz Inequality, we see that A is well posed, i.e. for any $u \in H_0^1(0, 1)$, it defines a linear and a continuous functional. Indeed, for a fixed $u \in H_0^1(0, 1)$, we have

$$\langle A(u), v \rangle \leq \|u\|_{H_0^1} \|v\|_{H_0^1} \text{ for all } v \in H_0^1(0, 1).$$

It now follows that A is continuous. We see that A is strongly monotone since for all $u, v \in H_0^1(0, 1)$, it holds

$$\langle A(u) - A(v), u - v \rangle = (u - v, u - v)_{H_0^1} =$$
$$\int_0^1 (\dot{u}(t) - \dot{v}(t)) (\dot{u}(t) - \dot{v}(t)) \, dt = \|u - v\|_{H_0^1}^2 \, .$$

Operator A is called the (negative) Laplacian and is denoted by $-\Delta$ or by $-\frac{d^2}{dt^2}$. This operator corresponds to our introductory problem (1.1) for $a = 0$.

Exercise 3.1 Let $m > 0$. Consider

$$A_0 : L^2(0, 1) \to L^2(0, 1)$$

defined by

$$A_0(u(t)) = b(t)u(t) \text{ for a.e. } t \in [0, 1],$$

where

$$b \in L^\infty(0, 1), \ b(t) \geq m \text{ for a.e. } t \in [0, 1].$$

Show that A_0 is m-strongly monotone.

Observe that an easy calculation demonstrates that a function $f : \mathbb{R} \to \mathbb{R}$ given by formula $f(x) = x + \frac{1}{2}\sin x$ is monotone. This can be generalized as follows:

Proposition 3.1
Assume that E is a real Hilbert space and that $T : E \to E$ is a contraction, i.e. there is a constant $0 < \alpha < 1$ such that for all $u, v \in E$,

$$\|T(u) - T(v)\| \leq \alpha \|u - v\|. \tag{3.2}$$

Then both I (the identity) and $A = I - T$ are strongly monotone.

Proof The strong monotonicity of I is obvious. Now, we turn to the examination of an operator $I - T$. Indeed, for any $u, v \in E$, we have

$$
\begin{aligned}
(A(u) - A(v), u - v)_E &= (u - T(u) - (v - T(v)), u - v)_E = \\
(u - v, u - v)_E &- (T(u) - (T(v)), u - v)_E \geq \\
(u - v, u - v)_E &- \|T(u) - T(v)\| \|u - v\| \geq (1 - \alpha) \|u - v\|^2.
\end{aligned}
$$

Remark 3.1
When T is nonexpansive, i.e. $\alpha = 1$ in (3.2), then A is monotone.

Exercise 3.2 Let $p \geq 2$. Using the Hölder Inequality, prove that formula

$$\langle A(u), v \rangle_{L^q, L^p} = \int_0^1 |u(t)|^{p-2} u(t) v(t) \, dt \tag{3.3}$$

for $u, v \in L^p (0, 1)$ defines operator from $L^p (0, 1)$ to $L^q (0, 1)$. Note that operator A can be defined pointwise on $[0, 1]$ by

$$A (u (\cdot)) = |u (\cdot)|^{p-2} u (\cdot).$$

Exercise 3.3 Let $p \geq 2$. Use the well known inequality

$$\left(|a|^{p-2} a - |b|^{p-2} b \right) (a - b) \geq 2^{2-p} |a - b|^p \text{ for } a, b \in \mathbb{R}^N$$

in order to show that operator A defined by (3.3) is uniformly monotone.

Remark 3.2

  (i) For a uniformly monotone operator, we have the following estimation for any $u \in E$:

$$\langle A (u), u \rangle \geq \|u\| (\rho (\|u\|) - \|A (0)\|_*).$$

 (ii) A d-monotone operator is monotone but not necessarily strictly monotone. Indeed, since ρ is increasing, it follows for all $u, v \in E$ that

$$(\rho (\|u\|) - \rho (\|v\|)) (\|u\| - \|v\|) \geq 0.$$

(iii) Taking $\rho (x) = mx$, we see that a strongly monotone operator is uniformly monotone.

(iv) A strongly monotone operator is obviously d-monotone with $\rho (x) = mx$.

 (v) When E is strictly convex, then a d-monotone operator is strictly monotone. Indeed, for any $u, v \in E$ such that

$$\begin{aligned}
0 = \langle A (u) - A (v), u - v \rangle = \\
2 \langle A (u) - A \left(\tfrac{u+v}{2} \right), u - \tfrac{u+v}{2} \rangle + 2 \langle A \left(\tfrac{u+v}{2} \right) - A (v), \tfrac{u+v}{2} - v \rangle \geq \\
2 \left(\rho (\|u\|) - \rho \left(\left\| \tfrac{u+v}{2} \right\| \right) \right) \left(\|u\| - \left\| \tfrac{u+v}{2} \right\| \right) + \\
2 \left(\rho \left(\left\| \tfrac{u+v}{2} \right\| \right) - \rho (\|v\|) \right) \left(\left\| \tfrac{u+v}{2} \right\| - \|v\| \right).
\end{aligned}$$

Then we see that

$$\left\| \frac{u + v}{2} \right\| = \|v\| = \|u\|,$$

and since E is strictly convex, we have $u = v$.

Exercise 3.4 Verify relations (i)–(iv) in the above remark.

Exercise 3.5 Let E be a Hilbert space. Show that if a bounded linear operator $A : E \to E$ is skew, that is, $A^* = -A$, then it is monotone.

Exercise 3.6 Show that the sum of two operators of various types of monotony forms an operator of the stronger type of monotony.

3.2 On Some Properties of Monotone Operators

There is a direct connection between the convexity of $F : E \to \mathbb{R}$ and the monotonicity of its derivative as seen below:

Theorem 3.1 (Convexity vs. Monotonicity)
Assume that $F : E \to \mathbb{R}$ is differentiable in the sense of Gâteaux. Then the following are equivalent:

(i) functional F is convex;
(ii) for all $u, v \in E$ function $s \mapsto F(u + sv)$ is convex on $\mathbb{R}$;
(iii) operator $F' : E \to E^*$ is monotone; and
(iv) for all $u, v \in E$ we have

$$F(u) - F(v) \geq \left\langle F'(v), (u - v) \right\rangle.$$

Proof We know from Theorem 2.16 that (iv) implies (i), and from (i), it easily follows that (ii) holds. It suffices to demonstrate that (ii) implies (iii) and that (iii) implies (iv).

We show that (ii) implies (iii). Fix $u, v \in E$. Put $g : [0, 1] \to \mathbb{R}$ by

$$g(s) = F(u + sv).$$

Function g is convex and differentiable on $[0, 1]$, and thus its derivative

$$g'(s) = \left\langle F'(u + sv), v \right\rangle$$

is nondecreasing on $[0, 1]$. By (1.6), this means that operator $F' : E \to E^*$ is monotone.

We now show that (iii) implies (iv). Fix $u, v \in E$ and consider $g : [0, 1] \to \mathbb{R}$ given by

$$g(s) = F(u + s(v - u)),$$

which is differentiable on $[0, 1]$. Its derivative

$$g^{'}(s) = \left\langle F^{'}(u + s(v - u)), v - u \right\rangle$$

is nondecreasing on $[0, 1]$. Thus

$$\left\langle F^{'}(u + s(v - u)), v - u \right\rangle \geq \left\langle F^{'}(u), v - u \right\rangle \text{ for } s \in [0, 1].$$

Since

$$g(1) - g(0) = \int_0^1 g^{'}(s)\,ds,$$

we obtain that

$$F(v) - F(u) = \int_0^1 \left\langle F^{'}(u + s(v - u)), v - u \right\rangle ds \geq$$
$$\int_0^1 \left\langle F^{'}(u), v - u \right\rangle ds = \left\langle F^{'}(u), v - u \right\rangle.$$

Therefore the theorem is now proved.

Sufficient condition for an operator to be monotone is as follows:

Proposition 3.2 (Sufficient Conditions for Monotonicity)
Operator $A : E \to E^*$ is monotone if and only if for any fixed $u, v \in E$, function $g : [0, 1] \to \mathbb{R}$ given by

$$g(s) = \langle A(u + sv), v \rangle$$

is nondecreasing. If additionally A has a Gâteaux derivative at every point such that for any $z, w \in E$, function

$$t \to \left\langle A^{'}(z + tw)w, w \right\rangle$$

is continuous on $[0, 1]$, then A is monotone if and only if for any $u, v \in E$ it holds

$$\left\langle A^{'}(u)v, v \right\rangle \geq 0. \tag{3.4}$$

Proof Assume that A is monotone. Let $t, s \in [0, 1]$, $t \le s$. Then for all $u, v \in E$,

$$g(s) - g(t) = \langle A(u + sv), v \rangle - \langle A(u + tv), v \rangle =$$
$$\frac{1}{s-t} \langle A(u + sv) - A(u + tv), (u + sv) - (u + tv) \rangle \ge 0.$$

This means that g is nondecreasing.

Assume now that g is nondecreasing, and put $s = 1$ and $t = 0$ in the above. It easily follows for all $u, v \in E$ that

$$\langle A(u + v) - A(u), v \rangle \ge 0.$$

Therefore we see that A is monotone.

Now let A be Gâteaux differentiable satisfying the above mentioned continuity assumption. Let A be monotone. Take $s \in (0, 1)$ and fix $u, v \in E$. We have

$$0 \le \langle A(u + sv) - A(u), sv \rangle = \int_0^s \left\langle A'(u + tv) v, sv \right\rangle dt.$$

Using the Integral Mean Value Theorem, we see that there is $s_0 \in [0, s]$ such that

$$\left\langle A'(u + s_0 v) v, sv \right\rangle = \frac{1}{s} \int_0^s \left\langle A'(u + tv) v, sv \right\rangle dt.$$

This means that

$$0 \le \langle A(u + sv) - A(u), sv \rangle = s^2 \left\langle A'(u + s_0 v) v, v \right\rangle,$$

and we see letting $s \to 0$ that (3.4) satisfied.

Now using assumption (3.4) and reversing the above arguments, we can conclude the proof by obtaining the remaining assertion.

Exercise 3.7   Assume that $A : E \to E^*$ has a Gâteaux derivative at every point and that for any fixed $u, w \in E$, function

$$t \to \left\langle A'(u + tw) w, w \right\rangle$$

is continuous on $[0, 1]$. Prove that A is strongly monotone if and only if there is a constant $m > 0$ such that for any $u, v \in E$ it holds

$$\left\langle A'(u) v, v \right\rangle \ge m \|v\|^2.$$

In what follows we use the Banach–Steinhaus Theorem (proved as Theorem 2.1.4 in [14] via the uniform boundedness principle) in order to examine some further properties of a monotone operator.

Theorem 3.2 (Banach–Steinhaus Theorem)

Assume that X is a Banach space and Y is a normed space, and consider a sequence $(A_n)_{n=1}^\infty \subset \mathscr{L}(X, Y)$. If for each $x \in X$, the sequence $(A_n(x))_{n=1}^\infty$ is bounded, then also the sequence $(\|A_n\|)_{n=1}^\infty$ is bounded, where for $n \in \mathbb{N}$,

$$\|A_n\| = \sup_{\|x\|=1} \|A_n x\|.$$

Definition 3.2 (Locally Bounded Operator)

Operator $A : E \to E^*$ is called locally bounded if for each fixed $u \in E$, there are constants $\varepsilon > 0$ and $M > 0$ such that it holds $\|A(v)\|_* \le M$ for all v satisfying $\|u - v\| \le \varepsilon$.

Proposition 3.3

Let $A : E \to E^*$ be a monotone operator. Then A is locally bounded.

Proof Suppose that A is not locally bounded. Then there exist a sequence $(u_n)_{n=1}^\infty \subset E$ and an element u_0 such that $u_n \to u_0$ in E and also that

$$\|A(u_n)\|_* \to +\infty.$$

For $n = 1, 2, \dots$, put

$$\alpha_n = 1 + \|A(u_n)\|_* \|u_n - u_0\|.$$

Since A is monotone, it follows for any $v \in E$

$$\langle A(u_n) - A(u_0 + v), u_n - (u_0 + v)\rangle \ge 0.$$

By definition of a sequence $(\alpha_n)_{n=1}^\infty$, there exists a constant M_1 such that for all $n \in \mathbb{N}$,

$$\begin{aligned}
\tfrac{1}{\alpha_n} \langle A(u_n), v\rangle &\le \\
\tfrac{1}{\alpha_n} (\langle A(u_n), u_n - u_0\rangle - \langle A(u_0 + v), u_n - (u_0 + v)\rangle) &\le \\
1 + \tfrac{1}{\alpha_n} \|A(u_0 + v)\|_* (\|u_n - u_0\| + \|v\|) &\le M_1 := M_1(u_0, v).
\end{aligned}$$

Since the above holds for $-v$ as well, we have

$$\limsup_{n \to +\infty} \left| \frac{1}{\alpha_n} \langle A(u_n), v\rangle \right| < +\infty$$

for all $v \in E$. Then by the Banach–Steinhaus Theorem, we see that

$$\frac{1}{\alpha_n} \|A(u_n)\|_* \leq M_2$$

for some constant M_2. Therefore

$$\|A(u_n)\|_* \leq \alpha_n M_2 = M_2 + M_2 \|A(u_n)\|_* \|u_n - u_0\|.$$

Choosing n_0 so that for $n \geq n_0$ it holds

$$M_2 \|u_n - u_0\| \leq \frac{1}{2},$$

we see that

$$\|A(u_n)\|_* \leq 2M_2,$$

which contradicts

$$\|A(u_n)\|_* \to +\infty.$$

There is also a much more demanding notion of a bounded operator.

Definition 3.3 (Bounded Operator)
Operator $A : E \to E^*$ is called bounded when for a bounded set $B \subset E$, set $A(B) \subset E^*$ is bounded as well.

Recall that a linear operator is bounded if and only if it is continuous. A (nonlinear) continuous operator need not be bounded as seen from the following example:

Example 3.3 Let us consider a continuous operator $A : l^2 \to l^2$ given by

$$A(u) = \left(u_1, u_2^2, u_3^4, \ldots\right).$$

Put

$$e_n = (\delta_{in})_{i=1}^{+\infty}, \quad \text{for } n \in \mathbb{N},$$

where

$$\delta_{in} = \begin{cases} 1, & i = n, \\ 0, & i \neq n \end{cases}$$

for $i, n \in \mathbb{N}$. For a bounded sequence $(2e_n) \subset l^2$, we have $\|2e_n\| = 2$ and $\|A(2e_n)\| = 2^n$. Thus A is not bounded.

Exercise 3.8 Show that operator A defined by (3.3) is bounded.

The Krasnosel'skii Theorem 2.12 helps us in proving that a given nonlinear operator is bounded. Note that a monotone operator need not be bounded. As an exercise, the reader should verify the details of an example which now follows.

Example 3.4 We consider a 1-strongly monotone operator $A : l^2 \to l^2$ given by the formula

$$A(x) = (f_1(x_1), f_2(x_2), \dots,) + (x_1, x_2, \dots,),$$

where

$$f_n(t) = \begin{cases} 0, & \text{for } t \le \frac{1}{2}, \\ nt - \frac{n}{2}, & \text{for } t > \frac{1}{2}. \end{cases}$$

Then $A\left(\overline{B(0, 1)}\right)$ is an unbounded set.

Now we introduce some tool that simplifies checking both the monotonicity and the boundedness of a nonlinear operator, see Theorem 4.10 from [53]:

Lemma 3.1

Let X be a real Banach space, and let $A \in \mathcal{L}(E, X)$. Then there exists exactly one operator

$$A^* \in \mathcal{L}(X^*, E^*)$$

such that for any $x \in E$, $y^* \in X^*$ it holds

$$\langle y^*, Ax \rangle_{X^*, X} = \langle A^* y^*, x \rangle_{E^*, E}.$$

Operator A^* is called **the adjoint** to A.

Example 3.5 When $A : L^2(0, 1) \to L^2(0, 1)$ is an identity, then $A^* = A$. Moreover, if $A : H_0^1(0, 1) \to L^2(0, 1)$ is given by $Au(\cdot) = \dot{u}(\cdot)$, then for $u, v \in H_0^1(0, 1)$ it holds

$$\langle A^* u, v \rangle = \int_0^1 u(t) \dot{v}(t) \, dt = \int_0^1 (-\dot{u}(t)) v(t) \, dt.$$

Lemma 3.2

Assume X is a real Banach space. Let $L : E \to X$ be such a linear operator that for $x \in E$ it holds

$$\|x\|_E = \|Lx\|_X . \tag{3.5}$$

Assume $A : X \to X^*$ has any monotonicity property (namely, A is monotone or strictly monotone or strongly monotone or uniformly monotone or d-monotone). Then operator $T : E \to E^*$ defined as follows

$$T = L^* A L$$

shares the monotonicity property of A.

Proof Suppose that A is monotone. Then for any $x, y \in E$, using the definition of the adjoint operator,

$$0 \leq \langle A(Lx) - A(Ly), Lx - Ly \rangle_{X^*, X} = \langle L^* A(Lx) - L^* A(Ly), x - y \rangle_{E^*, E} .$$

Remaining assertions follow likewise.

Exercise 3.9 Prove the remaining assertions in the above theorem.

Exercise 3.10 Check if (3.5) can be replaced with

$$\|Lx\|_X \geq \|x\|_E \quad \text{for all } x \in E.$$

Remark 3.3

Lemma 3.2 may also help in proving that operator is bounded. In fact in typical situation when we consider operator $T = L^* A L$, it is the case that A acts between Lebesgue integrable spaces and its boundedness follows by conditions used in Theorem 2.12 or else Theorem 2.13.

3.3 Different Types of Continuity

For nonlinear operators, there are a number of continuity notions. We provide only those that will be useful in the sequel.

Definition 3.4 (Different Types of Continuity)
Operator $A : E \to E^*$ is called

(i) radially continuous, if for all $u, v \in E$, function

$$s \to \langle A(u + sv), v \rangle$$

   is continuous on $[0, 1]$;
(ii) hemicontinuous, if for all $u, v, h \in E$, function

$$s \to \langle A(u + sv), h \rangle$$

   is continuous on $[0, 1]$;
(iii) demicontinuous if $u_n \to u_0$ in E implies that $A(u_n) \rightharpoonup A(u_0)$ in E^*;
(iv) Lipschitz continuous, if there exists a constant $L > 0$ such that

$$\|A(u) - A(v)\|_* \leq L \|u - v\|$$

   for all $u, v \in E$;
(v) uniformly continuous, if there exists a nondecreasing continuous function
   $\mu : [0, +\infty) \to [0, +\infty)$ such that for all $u, v \in E$ it holds

$$\|A(u) - A(v)\|_* \leq \mu(\|u - v\|);$$

(vi) strongly continuous if $u_n \rightharpoonup u_0$ in E implies that $A(u_n) \to A(u_0)$ in E^*;
(vii) weakly continuous if $u_n \rightharpoonup u_0$ in E implies that $A(u_n) \rightharpoonup A(u_0)$ in E^*.

Strong continuity that will be used often further on can be checked with the help of a compact embedding (i.e. Theorem 2.9) and the Krasnosel'skii Theorem. Now a number of exercises follow which allow for better understanding of the above introduced notions and serve as examples as well.

Exercise 3.11 Find a function $f : \mathbb{R}^2 \to \mathbb{R}^2$ that is hemicontinuous at $(0, 0)$ but not continuous. Hint: consider functions continuous on lines only.

Exercise 3.12 Let $A : E \to E^*$ be weakly continuous. Show that it is demicontinuous.

Exercise 3.13 Let E be a finite dimensional Banach space and $A : E \to E^*$ be a monotone and hemicontinuous operator. Show that A is continuous. Is this assertion true for a non-monotone mapping?

Exercise 3.14 Let $A : E \to E^*$ be strongly continuous. Show that it is compact (i.e. A is continuous and sends bounded sets into relatively compact ones).

Exercise 3.15 Let $A : E \to E^*$ be Lipschitz continuous. Show that it is uniformly continuous.

Exercise 3.16 Show that a uniformly continuous operator is bounded.

Exercise 3.17 Assume that $T : E \to E^*$ is a strongly monotone operator and that $A : E \to E^*$ is Lipschitz continuous with respect to a constant $L > 0$. Find values of α for which operator $T + \alpha A$ is

(i) monotone and
(ii) strongly monotone.

Exercise 3.18 Show that if $A : E \to E^*$ is demicontinuous, then it is hemicontinuous and locally bounded. (Observe that monotonicity is not assumed.)

Exercise 3.19 Show that if $A : E \to E^*$ is linear and demicontinuous, then it is continuous.

Exercise 3.20 Let $A : l^2 \to l^2$ be defined by $A(u) = (\|u\|, 0, 0, ...)$. Show that A is continuous and compact, but it is not strongly continuous.

Proposition 3.4
A monotone and linear operator $A : E \to E^*$ is continuous.

Proof Let $u_n \to u_0$ in E. Put for $n = 1, 2, ...,$

$$v_n = \begin{cases} \frac{u_n - u_0}{\sqrt{\|u_n - u_0\|}}, & \text{when } u_n \neq u_0 \\ 0, & \text{when } u_n = u_0. \end{cases}$$

Then $v_n \to 0$. Since A is locally bounded it follows that there is $M > 0$ such that $\|A(v_n)\|_* \leq M$ for $n \in \mathbb{N}$. Then we have for $n \in \mathbb{N}$

$$\|A(u_n) - A(u_0)\|_* = \|A(u_n - u_0)\|_* =$$
$$\sqrt{\|u_n - u_0\|} \, \|A(v_n)\|_* \leq M\sqrt{\|u_n - u_0\|}.$$

The above means that $\|A(u_n) - A(u_0)\|_* \to 0$ as $n \to +\infty$.

To the end of this section, we give the following example of a strongly continuous operator connected to Example 2.12.

Example 3.6 Assume that $f : [0, 1] \times \mathbb{R}^N \to \mathbb{R}^N$ is a Carathéodory function and there exist

$$\phi \in L^\infty (0, 1), \; h \in L^1 (0, 1)$$

such that

$$|f (t, x)| \leq \phi (t) |x|^2 + h (t)$$

for a.e. $t \in [0, 1]$ and all $x \in \mathbb{R}^N$. Define operator $A : H_0^1 (0, 1) \to H^{-1} (0, 1)$ by

$$\langle A (u), v \rangle = \int_0^1 f (t, u (t)) v (t) \, dt \text{ for all } u, v \in H_0^1 (0, 1). \tag{3.6}$$

In Example 2.12, we proved that functional $J : H_0^1 (0, 1) \to \mathbb{R}$ defined by

$$J (u) = \int_0^1 f (t, u (t)) \, dt$$

is sequentially weakly continuous. Now, we prove that operator A defined above is strongly continuous. The arguments follow almost likewise. Observe firstly that A is well defined, i.e. $A (u)$ defines a continuous functional on $H_0^1 (0, 1)$ for any fixed $u \in H_0^1 (0, 1)$. Indeed, any $u \in H_0^1 (0, 1)$ is continuous and thus bounded by some $d_u > 0$. This means that for a.e. $t \in [0, 1]$, we have

$$f (t, u (t)) \leq \phi (t) d_u^2 + h (t).$$

Then function $h_u (\cdot) = \phi (\cdot) d_u^2 + h (\cdot)$ is an element of $L^1 (0, 1)$. We see that for all $v \in H_0^1 (0, 1)$ by the Sobolev Inequality, it holds

$$\left| \int_0^1 f (t, u (t)) v (t) \, dt \right| \leq \|h_u\|_{L^1} \|v\|_C \leq \|h_u\|_{L^1} \|v\|_{H_0^1}.$$

This proves that A is well defined. Now, take a sequence $(u_n)_{n=1}^\infty$, which is weakly convergent to some u_0 in $H_0^1 (0, 1)$. Then $(u_n)_{n=1}^\infty$ converges uniformly on $[0, 1]$ to u_0. Thus there is some $d > 0$ that $\|u_n\|_C \leq d$. This further implies that

$$|f (t, u_n (t))| \leq d^2 |\phi (t)| + |h (t)|$$

for a.e. $t \in [0, 1]$. Hence applying the Lebesgue Dominated Convergence Theorem, we obtain the assertion of the example. The operator discussed here corresponds to (1.1) with a nonlinear term f.

Exercise 3.21 Assume that $f : [0, 1] \times \mathbb{R}^N \to \mathbb{R}^N$ is a Carathéodory function, and there exist $\phi \in L^2(0, 1)$ and $h \in L^1(0, 1)$ such that

$$|f(t, x)| \leq \phi(t)|x| + h(t)$$

for a.e. $t \in [0, 1]$ and all $x \in \mathbb{R}^N$. Prove that operator A given by (3.6) is strongly continuous.

3.4 Coercivity

Coercivity is a notion which is aimed at having the set of solutions to equation $A(u) = b$ bounded. We have already introduced this notion in a finite dimensional setting. What follows a suitably modified definition:

Definition 3.5 (Coercive Operator)
Operator $A : E \to E^*$ is weakly coercive when

$$\lim_{\|u\| \to +\infty} \|A(u)\|_* = +\infty.$$

Operator A is called coercive if

$$\lim_{\|u\| \to +\infty} \frac{\langle A(u), u \rangle}{\|u\|} = +\infty.$$

Exercise 3.22 Prove that if there exists a function

$$\gamma : [0, +\infty) \to \mathbb{R}, \quad \lim_{x \to +\infty} \gamma(x) = +\infty,$$

such that

$$\langle A(u), u \rangle \geq \gamma(\|u\|)\|u\| \quad \text{for all } u \in E,$$

then $A : E \to E^*$ is coercive.

Remark 3.4
By adding a monotone operator of any kind to a coercive operator, we obtain a coercive operator as well.

As observed from Example 3.4, we see that a coercive operator need not be bounded. Now a number of exercises follow providing examples of coercive operators as well.

Exercise 3.23 Show that a coercive operator is weakly coercive. Show that if $A : E \to E^*$ is coercive and if $T : E \to E^*$ is such that

$$\langle T(v), v \rangle \geq c \text{ for all } v \in E,$$

and some fixed constant $c \in \mathbb{R}$, then $A + T$ is coercive.

Exercise 3.24 Show that a strongly monotone operator is coercive.

Exercise 3.25 Show that a d-monotone operator is coercive when $\rho(x) \to +\infty$ as $x \to +\infty$; here, ρ is a function from (3.1).

Exercise 3.26 Assume that $A : E \to E^*$ is uniformly monotone and such that $A(0) = 0$. Prove that A is coercive.

Example 3.7 Let $a \in \mathbb{R}$. Define operator

$$A : H_0^1(0, 1) \to H^{-1}(0, 1)$$

by

$$\langle A(u), v \rangle = \int_0^1 \dot{u}(t) \dot{v}(t) \, dt + a \int_0^1 \dot{u}(t) v(t) \, dt \text{ for } u, v \in E.$$

This operator corresponds to our introductory problem (1.1). Observe that for any $a \in \mathbb{R}$, operator A is coercive. Indeed, for $u \in H_0^1(0, 1)$, it holds

$$\int_0^1 \dot{u}(t) u(t) \, dt = \int_0^1 \left(\frac{1}{2} |u(t)|^2 \right)' dt = \frac{1}{2} |u(1)|^2 - \frac{1}{2} |u(0)|^2 = 0.$$

Thus for $u \in H_0^1(0, 1)$, we have

$$\langle A(u), u \rangle = \|u\|_{H_0^1}^2,$$

which implies the coercivity of A.

Exercise 3.27 Using the Poincaré Inequality, find the values of parameter a for which the operator A from the above example is strongly monotone. Are there any values of $a \in \mathbb{R}$ for which A is merely monotone? Prove the assertion of Example 3.7 using Remark 3.4 for such values of $a \in \mathbb{R}$.

Remark 3.5

A uniformly monotone operator $A : E \to E^*$ is coercive. Indeed, take $u \in E, u \neq 0$, and define

$$v = \frac{u}{\|u\|}, \quad n = [\|u\|] .$$

Note that when $\|u\| > n$, it follows

$$\langle A(\|u\| v) - A(nv), v \rangle =$$
$$\frac{1}{\|u\|-n} \langle A(\|u\| v) - A(nv), \|u\| v - nv \rangle > 0.$$

Then we have

$$\langle A(u), u \rangle = \|u\| \langle A(u), v \rangle =$$
$$\|u\| \left(\langle A(\|u\| v) - A(nv), v \rangle + \langle A(nv) - A(0), v \rangle + \langle A(0), v \rangle \right) \geq$$
$$\|u\| \left(\langle A(nv) - A(0), v \rangle - \|A(0)\|_* \right) =$$
$$\|u\| \left(\sum_{j=1}^{n} \langle A(jv) - A((j-1)v), v \rangle - \|A(0)\|_* \right) \geq$$
$$\|u\| \left(n\rho(1) - \|A(0)\|_* \right) \geq \|u\| \left((\|u\| - 1) \rho(1) - \|A(0)\|_* \right).$$

Thus if we put

$$\gamma(x) = \rho(1) \cdot (x - 1) - \|A(0)\|_{E^*} \text{ for } x \geq 0,$$

we see that A is coercive.

3.5 An Example of a Monotone Mapping

In this section, we are concerned with an example of a nonlinear monotone mapping. We will introduce several assumptions leading to various types of monotone operators. Let $p \geq 2$. We assume that

Aφ $\varphi : [0, 1] \times \mathbb{R}_+ \to \mathbb{R}$ is a Carathéodory function for which there is a constant $M > 0$ such that

$$|\varphi(t, x)| \leq M \text{ for a.e. } t \in [0, 1] \text{ and all } x \in \mathbb{R}_+.$$

Under some additional growth assumption on function φ, we will consider the monotonicity of an operator

$$A : L^p(0, 1) \to L^q(0, 1),$$

given by

$$\langle A(u), v \rangle = \int_0^1 \varphi\left(t, |u(t)|^{p-1}\right) |u(t)|^{p-2} u(t) v(t) \, dt \tag{3.7}$$

or else a.e. on $[0, 1]$

$$A(u(\cdot)) = \varphi\left(\cdot, |u(\cdot)|^{p-1}\right) |u(\cdot)|^{p-2} u(\cdot). \tag{3.8}$$

Exercise 3.28 Verify by a direct calculation that A is well defined, and show that it is continuous. Hint: due to the bound on φ, the Krasnosel'skii Theorem 2.12 can be applied.

Theorem 3.3

Assume that condition $A\varphi$ is satisfied. The following assertions hold.

(i) *If*

$$\varphi(t, x) x - \varphi(t, y) y \geq 0$$

for all $x \geq y \geq 0$ and a.e. $t \in [0, 1]$, then A is monotone.

(ii) *If there exists a constant $\gamma > 0$ such that*

$$\varphi(t, x) x - \varphi(t, y) y \geq \gamma (x - y)$$

for all $x \geq y \geq 0$ and a.e. $t \in [0, 1]$, then A is d-monotone with respect to function

$$\rho(x) = \gamma x^{p-1}.$$

(iii) *If for some $m > 0$,*

$$\varphi(t, x) \geq m$$

for all $x \in \mathbb{R}_+$ and a.e. $t \in [0, 1]$, then A is coercive.

(iv) *Let $p = 2$ and let $L > 0$. If*

$$|\varphi(t, |x|) x - \varphi(t, |y|) y| \leq L |x - y| \tag{3.9}$$

for all $x, y \in \mathbb{R}$ and a.e. $t \in [0, 1]$, then A is Lipschitz continuous.

(continued)

Theorem 3.3 (continued)

  (v) Let $p = 2$, and let there exist a constant $\gamma > 0$ such that

$$\varphi(t, x)\, x - \varphi(t, y)\, y \geq \gamma\,(x - y) \tag{3.10}$$

  for all $x \geq y \geq 0$ and a.e. $t \in [0, 1]$. Then A is strongly monotone.

  (vi) Let $p \geq 2$. If a function $x \mapsto \varphi(t, x)$ is continuously differentiable for a.e.
      $t \in [0, 1]$ on $\mathbb{R}_+$ and if there exists a constant $M > 0$ such that

$$\left| \frac{\partial}{\partial x} \varphi(t, x)\, x \right| \leq M \text{ for all } x \in \mathbb{R}_+ \text{ and a.e. } t \in [0, 1]\,,$$

  then A is differentiable in the sense of Gâteaux.

Proof Note that in what follows we write sometimes u instead of $u(t)$ under the integral sign in order to shorten the notation. We divide the proof into several steps corresponding to assertions (i)–(vi).

(i) Observe that for $u, v \in L^p(0, 1)$,

$$\langle A(u) - A(v), u - v \rangle =$$
$$\int_0^1 \left(\varphi\left(t, |u|^{p-1}\right) |u|^{p-2} u - \varphi\left(t, |v|^{p-1}\right) |v|^{p-2} v \right) (u - v)\, dt =$$
$$\int_0^1 \varphi\left(t, |u|^{p-1}\right) |u|^{p-2} u\,(u - v)\, dt - \int_0^1 \varphi\left(t, |v|^{p-1}\right) |v|^{p-2} v\,(u - v)\, dt \geq$$
$$\int_0^1 \varphi\left(t, |u|^{p-1}\right) |u|^{p-2} \left(|u|^2 - |u|\,|v|\right) dt - \int_0^1 \varphi\left(t, |v|^{p-1}\right) |v|^{p-2} \left(|u|\,|v| - |v|^2\right) dt =$$
$$\int_0^1 \left(\varphi\left(t, |u|^{p-1}\right) |u|^{p-1} - \varphi\left(t, |v|^{p-1}\right) |v|^{p-1} \right) (|u| - |v|)\, dt \geq 0.$$

(ii) From (i) and by the Hölder Inequality, we see for $u, v \in L^p(0, 1)$

$$\langle A(u) - A(v), u - v \rangle \geq$$
$$\int_0^1 \left(\varphi\left(t, |u|^{p-1}\right) |u|^{p-1} - \varphi\left(t, |v|^{p-1}\right) |v|^{p-1} \right) (|u| - |v|)\, dt \geq$$
$$\gamma \int_0^1 \left(|u|^{p-1} - |v|^{p-1}\right) (|u| - |v|)\, dt =$$
$$\gamma \int_0^1 |u|^p\, dt + \gamma \int_0^1 |v|^p\, dt - \gamma \int_0^1 \left(|u|^{p-1} |v| - |v|^{p-1} |u|\right) dt \geq$$
$$\gamma \left(\|u\|_{L^p}^p + \|v\|_{L^p}^p - \|u\|_{L^p}^{p-1} \|v\|_{L^p} - \|u\|_{L^p} \|v\|_{L^p}^{p-1} \right) =$$
$$\gamma \left(\|u\|_{L^p} - \|v\|_{L^p} \right) \left(\|u\|_{L^p}^{p-1} - \|v\|_{L^p}^{p-1} \right).$$

(iii) Note that for any $u \in L^p(0, 1)$, we have

$$\langle A(u), u \rangle = \int_0^1 \varphi\left(t, |u|^{p-1}\right) |u|^p\, dt \geq m \int_0^1 |u(t)|^p\, dt = m\, \|u\|_{L^p}^p\,.$$

This estimation implies the coercivity.

(iv) Since $p = 2$ from (3.9), we have for $u, v, w \in L^2(0, 1)$

$$|\langle A(u) - A(v), w \rangle| \le \int_0^1 |\varphi(t, |u(t)|) u(t) - \varphi(t, |v(t)|) v(t)| \cdot |w(t)| \, dt \le$$
$$L \int_0^1 |u(t) - v(t)| |w(t)| \, dt \le L \|u - v\|_{L^2} \|w\|_{L^2}.$$

By the above estimation, it easily follows that

$$\|A(u) - A(v)\|_{L^2} \le L \|u - v\|_{L^2}$$

for all $u, v \in L^p(0, 1)$.

(v) Since (3.10) holds, we can define function φ_0 by

$$\varphi_0(u(\cdot)) = \varphi(\cdot, |u(\cdot)|) - \gamma \text{ for } u \in L^2(0, 1).$$

Denote by A_0 the operator defined by (3.7) with φ_0 instead of φ. Then $\varphi = \varphi_0 + \gamma$. We also put $\varphi_1(u(\cdot)) = \gamma$ for $u \in L^2(0, 1)$ and observe that operator

$$A_1 : L^2(0, 1) \to L^2(0, 1)$$

determined by φ_1 according to formula (3.7) with φ_1 instead of φ is strongly monotone. By part (i), it follows that operator A_0 is monotone. A sum of a strongly monotone and a monotone operator, that is, operator A, is necessarily strongly monotone.

(vi) Let us fix $u, v \in L^p(0, 1)$, and let us consider the following function $\Phi : \mathbb{R} \to \mathbb{R}$

$$\Phi(s) = \int_0^1 \varphi\left(t, |u(t) + s \cdot z(t)|^{p-1}\right) |u(t) + s \cdot z(t)|^{p-2} (u(t) + s \cdot z(t)) v(t) \, dt.$$
$$(3.11)$$

Using the Lebesgue Dominated Convergence Theorem, we can prove that function Φ is differentiable and that

$$\left\langle A'(u) z, v \right\rangle = \Phi'(s)\Big|_{s=0} =$$
$$\int_0^1 \left\{ \varphi\left(t, |u(t)|^{p-1}\right) \left((p-2) |u(t)|^{p-2} v(t) z(t) + |u(t)|^{p-2} v(t) z(t)\right) \right. \quad (3.12)$$
$$\left. (p-1) \frac{\partial \varphi}{\partial x}\left(t, |u(t)|^{p-1}\right) |u(t)|^{2p-3} v(t) z(t) \right\} dt.$$

Exercise 3.29 Let $u, z \in L^p(0, 1)$ be fixed. Prove that under assumptions of assertion (vi) of Theorem 3.3, formula (3.12) defines a linear and continuous mapping (of variable v).

Exercise 3.30 Prove that function $\Phi : \mathbb{R} \to \mathbb{R}$ given by (3.11) is differentiable and that formula (3.12) holds.

Exercise 3.31 Consider assertion (vi) of Theorem 3.3 for $p = 2$. Derive explicitly the counterpart of (3.12) in this case.

3.6   Condition (S) and Some Other Related Notions

Recall that in a uniformly convex space, relations $u_n \rightharpoonup u_0$ and $\|u_n\| \to \|u_0\|$ imply that $u_n \to u_0$. This relation suggests the following notion:

Definition 3.6 (Condition (S))

Operator $A : E \to E^*$ satisfies condition (S), or we say that A has property (S), if relations

$$u_n \rightharpoonup u_0 \text{ in } E$$

and

$$\langle A(u_n) - A(u_0), u_n - u_0 \rangle \to 0$$

imply that

$$u_n \to u_0 \text{ in } E.$$

Remark 3.6

Take the following operator $A : H_0^1(0, 1) \to H^{-1}(0, 1)$

$$\langle A(u), v \rangle = \int_0^1 (\dot{u}(t)) \dot{v}(t) \, dt - \int_0^1 2u(t) v(t) \, dt.$$

Then we can easily calculate using the Poincaré Inequality that A is not monotone. Nevertheless recalling that a weakly convergent sequence $(u_n)_{n=1}^\infty$ from $H_0^1(0, 1)$ is strongly convergent in $L^2(0, 1)$, we see that relation

$$\langle A(u_n) - A(u_0), u_n - u_0 \rangle \to 0$$

implies that

$$\int_0^1 (\dot{u}_n(t) - \dot{u}_0(t))^2 \, dt = \langle A(u_n) - A(u_0), u_n - u_0 \rangle + 2 \int_0^1 (u_n(t) - u_0(t))^2 \, dt \to 0.$$

Thus operator A satisfies condition (S).

Lemma 3.3

Assume that E is a uniformly convex space. Let $A : E \to E^*$ be d-monotone. Then A satisfies condition (S).

Proof Assume that $u_n \rightharpoonup u_0$ in E and that $\langle A(u_n) - A(u_0), u_n - u_0 \rangle \to 0$. This by definition of a d-monotone operator means that

$$(\rho\,(\|u_n\|) - \rho\,(\|u_0\|))\,(\|u_n\| - \|u_0\|) \to 0 \text{ as } n \to +\infty$$

for some increasing function ρ. Therefore $\|u_n\| \to \|u_0\|$ as $n \to +\infty$. Since E is uniformly convex, we obtain that $u_n \to u_0$.

Exercise 3.32 Check that the uniformly monotone operator satisfies condition (S).

We have the following simple criterion.

Lemma 3.4

Assume that operator $A : E \to E^*$ fulfills property (S) and that $T : E \to E^*$ is strongly continuous. Then $A + T$ also has property (S).

Proof When T is strongly continuous, we obtain for a sequence $(u_n)_{n=1}^{\infty} \subset E$ weakly convergent to some u_0 that

$$\langle T(u_n) - T(u_0), u - u_0 \rangle \to 0.$$

Thus the assertion follows by a direct calculation.

There is also another applicable condition called $(S)_2$ which follows from condition (S).

Definition 3.7 (Condition $(S)_2$)

Operator $A : E \to E^*$ satisfies condition $(S)_2$ if $u_n \rightharpoonup u_0$ in E and $A(u_n) \to A(u_0)$ in E^* imply that $u_n \to u_0$ in E.

We have also other types of similar conditions, i.e. making weakly convergent sequences, via a suitable action of an operator, strongly convergent. These are as follows:

Condition $(S)_+$: $u_n \rightharpoonup u_0$ in E and

$$\limsup_{n \to +\infty} \langle A(u_n) - A(u_0), u_n - u_0 \rangle \le 0$$

imply that $u_n \to u_0$ in E;

Condition $(S)_0$:

$$u_n \rightharpoonup u_0, \quad A(u_n) \rightharpoonup b, \quad \langle A(u_n), u_n \rangle \to \langle b, u_0 \rangle$$

imply that $u_n \to u_0$ in E.

Remark 3.7
In the definitions of conditions (S) and $(S)_+$, one can replace

$$\langle A(u_n) - A(u_0), u - u_0 \rangle \to 0$$

with

$$\langle A(u_n), u_n - u_0 \rangle \to 0$$

since when $u_n \rightharpoonup u_0$ in E we always have

$$\langle A(u_0), u_n - u_0 \rangle \to 0.$$

Remark 3.8
In a recent paper [8], it is introduced another condition related to the above whose usage in the existence theory for nonlinear equations is then exploited.

Exercise 3.33 Prove that strongly continuous perturbations do not violate that operator A satisfies either of conditions $(S)_2$ or $(S)_+$ or else $(S)_0$. This means that the assertion of Lemma 3.4 holds with either of these conditions instead of (S).

Exercise 3.34 Prove that condition $(S)_+$ implies that condition (S) is satisfied and this in turn implies condition $(S)_0$.

3.7   The Minty Lemma and the Fundamental Lemma for Monotone Operators

We start with the following lemma about the monotonicity revealing one of its astonishing properties. The proof is based on what is now called the "Minty trick."

Lemma 3.5 (Minty Lemma)
Assume that operator $A : E \to E^*$ is monotone and hemicontinuous. Let $u_0 \in E$ be arbitrarily fixed. Then the following are equivalent:

(i) $\langle A(u_0), v - u_0 \rangle \geq 0$ for all $v \in E$ and
(ii) $\langle A(v), v - u_0 \rangle \geq 0$ for all $v \in E$.

Proof From the monotonicity of A, we obtain for all $v \in E$

$$\langle A(v), v - u_0 \rangle \geq \langle A(u_0), v - u_0 \rangle;$$

thus (i) implies (ii).

Assume that (ii) holds and take $v = u_0 + t(z - u_0)$ for any $t > 0$ and any $z \in E$. Then we have from (ii)

$$\langle A(u_0 + t(z - u_0)), z - u_0 \rangle \geq 0.$$

Since A is hemicontinuous, we see letting $t \to 0$ in the above that (ii) implies (i).

Definition 3.8

Let $D(A) \subset E$. Operator $A : D(A) \to E^*$ is called maximal monotone if relation

$$\langle f - A(v), u_0 - v \rangle \geq 0,$$

for all $v \in D(A)$, implies that $A(u_0) = f$.

Exercise 3.35 Assume that E is a Hilbert space and $A : E \to E$ is a linear and bounded mapping such that $(Ax, x)_E \geq 0$ for all $x \in E$. Prove that A is maximal monotone.

The notion of maximal monotonicity of A is equivalent to saying that the graph of operator A is not properly included to the graph of another monotone map. This means that the graph "has no gaps." Hence any type of continuity should help in having this condition satisfied. We have the result connecting the continuity notions for monotone operators as well as connections to the idea of the maximal monotonicity.

Lemma 3.6   (Fundamental Lemma for Monotone Operators)
Assume that operator $A : E \to E^*$ is monotone. Then the following are equivalent:

   (i) operator A is radially continuous;
   (ii) relation

$$\langle f - A(v), u_0 - v \rangle \geq 0$$

   for all $v \in E$, implies that $A(u_0) = f$;

(continued)

Lemma 3.6 (continued)

(iii) relations $u_n \rightharpoonup u_0$ in E, $A(u_n) \rightharpoonup f$ in E^* and

$$\limsup_{n \to +\infty} \langle A(u_n), u_n \rangle \leq \langle f, u_0 \rangle$$

imply that $A(u_0) = f$; and

(iv) operator A is demicontinuous.

Proof (i) implies (ii). Let $v \in E$ be arbitrary and define $v_t = u_0 - tv$ for $t > 0$. Then we have

$$t \langle f - A(v_t), v \rangle \geq 0.$$

Since A is radially continuous, we see taking $t \to 0$ that

$$\langle f - A(u_0), v \rangle \geq 0.$$

Repeating the above arguments for $v_t = u_0 + tv$, we also have that

$$\langle f - A(u_0), v \rangle = 0.$$

Since v is arbitrary, we observe that $A(u_0) = f$.

(ii) implies (iii). Assume that

$$u_n \rightharpoonup u_0 \text{ in } E, \quad A(u_n) \rightharpoonup f \text{ in } E^*$$

and

$$\limsup_{n \to +\infty} \langle A(u_n), u_n \rangle \leq \langle f, u_0 \rangle.$$

Then for any $v \in E$, we have

$$\langle f - A(v), u_0 - v \rangle = \langle f, u_0 \rangle - \langle f, v \rangle - \langle A(v), u_0 - v \rangle \geq$$
$$\limsup_{n \to +\infty} \langle A(u_n), u_n \rangle - \lim_{n \to +\infty} \langle A(u_n), v \rangle - \lim_{n \to +\infty} \langle A(v), u_n - v \rangle =$$
$$\limsup_{n \to +\infty} \langle A(u_n) - A(v), u_n - v \rangle \geq 0.$$

Thus (ii) implies that $A(u_0) = f$.

(iii) implies (iv). Let $u_n \to u_0$ in E. Since A is monotone, it is also locally bounded by Proposition 3.3. It follows that sequence $(A(u_n))_{n=1}^{\infty}$ has a weakly convergent subsequence, $\left(A\left(u_{n_k}\right)\right)_{k=1}^{\infty}$, which converges to some f. Since $u_{n_k} \to u_0$ in E as well, we see

$$\lim_{k \to +\infty} \left\langle A\left(u_{n_k}\right), u_{n_k} \right\rangle = \langle f, u_0 \rangle.$$

But this means that $A(u_0) = f$ and $A(u_{n_k}) \rightharpoonup A(u_0)$. Since any weakly convergent subsequence of $(A(u_n))_{n=1}^{\infty}$ converges to $A(u_0)$, we see that $A(u_n) \rightharpoonup A(u_0)$ and therefore A is demicontinuous.

(iv) implies (i). Since A is demicontinuous, it is also radially continuous.

Corollary 3.1

Assume that operator $A : E \to E^*$ is monotone. Then the following is equivalent to (i)–(iv) from Lemma 3.6:

(v) let $K \subset E$ be any dense subset, relation

$$\langle f - A(v), u_0 - v \rangle \geq 0,$$

for all $v \in K$, implies that $A(u_0) = f$.

Proof Since condition (iii) obviously implies condition (v), it suffice to show that

$$\langle f - A(v), u_0 - v \rangle \geq 0 \text{ for all } v \in K, \tag{3.13}$$

implies that

$$\langle f - A(v), u_0 - v \rangle \geq 0 \text{ for all } v \in E. \tag{3.14}$$

For any $v \in E$, there is a sequence $(v_n) \subset K$ such that $v_n \to v$. Then we have from (3.13)

$$\langle f - A(v_n), u_0 - v_n \rangle \geq 0.$$

Since A is demicontinuous, it follows that $A(v_n) \rightharpoonup A(v)$. Therefore, taking a limit, we see that (3.14) holds. Then, by (ii), we get the assertion.

Exercise 3.36 Show that conditions in Lemma 3.6 are equivalent to the following:
(vi) operator A is hemicontinuous.

From Lemma 3.6, it follows that a radially continuous (or demicontinuous or hemicontinuous) and monotone operator is maximal monotone. Since we usually assume monotonicity and some continuity about the operator, this is why we consider only maximal monotone operators. We will not comment on this in the sequel.

On the Fenchel-Young Conjugate

4

The results collected here will help us in understanding the issues related to the invertibility of nonlinear mappings. If we recall that a derivative of a differentiable convex functional is a monotone mapping, there is no surprise that we mainly work with convex functionals here. In order to introduce the Fenchel-Young conjugacy we used [15, 27].

4.1   Some Background from Convex Analysis

Let $F : E \to \mathbb{R} \cup \{+\infty\}$. The effective domain of F is the following set:

$$\mathrm{dom}\,(F) = \{x \in E : F(x) < +\infty\}.$$

When $\mathrm{dom}\,(F) \neq \emptyset$ we say that F is proper. We will consider only proper functionals in what follows. The reason why we will not consider functionals taking values $-\infty$ is as that a lower semicontinuous convex functional taking value $-\infty$ at one point cannot take any finite value (see Proposition 2.4, [15]). We have already considered the convexity of finite functionals (i.e. taking real values) but this is not sufficient for the purposes of our further investigations. The convexity and the strict convexity of $F : E \to \mathbb{R} \cup \{+\infty\}$ are defined in a standard manner according to formulas (2.19) and (2.20).

> **Theorem 4.1**
> *Functional $F : E \to \mathbb{R} \cup \{+\infty\}$ is convex if and only if the set $\mathrm{Epi}(F)$ is convex.*

Proof Let (x, α_x), $(y, \alpha_y) \in \text{Epi}\,(F)$ and let $\lambda \in [0, 1]$. We see that

$$\lambda \alpha_x + (1 - \lambda)\alpha_y \geq \lambda F(x) + (1 - \lambda)F(y) \geq F(\lambda x + (1 - \lambda)y).$$

Hence Epi (F) is convex.

Now let Epi (F) be convex and let $(x, F(x))$, $(y, F(y)) \in \text{Epi}\,(F)$ and let $\lambda \in [0, 1]$. Then

$$(\lambda x + (1 - \lambda)y, \lambda F(x) + (1 - \lambda)F(y)) \in \text{Epi}\,(F)$$

which implies that

$$\lambda F(x) + (1 - \lambda)F(y) \geq F(\lambda x + (1 - \lambda)y).$$

Although convexity is a geometric notion it has a great impact on continuity.

Theorem 4.2 (Continuity of a Convex Functional)
Let $C \subset E$ be open and convex and let $F : C \to \mathbb{R}$ be convex and bounded from the above in some neighborhood of $x_0 \in C$. Then F is continuous at x_0.

Proof We can always reduce the proof to the case of checking the continuity at $\overline{x} = 0$ provided that $F(\overline{x}) = 0$. We take an open ball $B(0, r)$ centered at 0 with radius r. Assume that there is a constant $M > 0$ such that

$$F(v) \leq M \text{ for all } v \in B(0, r).$$

Fix $\varepsilon \in (0, 1)$ and take any $x \in B(0, \varepsilon r)$. Observe that

$$\frac{x}{\varepsilon} \in B(0, r) \text{ and } -\frac{x}{\varepsilon} \in B(0, r).$$

It follows by the convexity of F that

$$F(x) = F\left((1 - \varepsilon)0 + \varepsilon\frac{x}{\varepsilon}\right) \leq (1 - \varepsilon)F(0) + \varepsilon F\left(\frac{x}{\varepsilon}\right) = \varepsilon F\left(\frac{x}{\varepsilon}\right) \leq \varepsilon M$$

and also

$$0 = (1 + \varepsilon)F(0) = (1 + \varepsilon)F\left(\frac{1}{1 + \varepsilon}x + \frac{\varepsilon}{1 + \varepsilon}\left(-\frac{x}{\varepsilon}\right)\right) \leq F(x) + \varepsilon F\left(-\frac{x}{\varepsilon}\right).$$

Summarizing for any $x \in B(0, \varepsilon r)$ we obtain the following estimation:

$$|F(x)| \leq \varepsilon M$$

which implies the continuity of F at $\overline{x} = 0$.

A simple corollary now follows:

Corollary 4.1
Every lower semicontinuous convex function over a Banach space is continuous over the interior of its effective domain.

The following exercises are to be treated as a word of caution when dealing with convex functions that are not defined on the whole space, i.e. which take value $+\infty$.

Exercise 4.1 Consider function $F : \mathbb{R}^2 \to \mathbb{R} \cup \{+\infty\}$ defined by

$$F(x, y) = \begin{cases} \frac{x^2}{y}, & y > 0, \\ 0, & (x, y) = 0, \\ +\infty, & \text{otherwise.} \end{cases}$$

Show that this function is convex and its epigraph is closed (i.e. it is lower semi continuous.)

Exercise 4.2 Show that function F defined above is not continuous at $(0, 0)$, i.e. F is continuous only in the interior of its effective domain.

From the above it is clear why we must impose the lower semicontinuity assumption on the convex function. In the sequel we shall need the separation theorem which is as follows (see Corollary 2.1.18 in [14]):

Theorem 4.3 (Separation Theorem)
Assume that C is closed and convex subset of E. Then for each $x \notin C$ there is a functional $x^* \in E^*$ such that

$$\langle x^*, x \rangle > \sup_{y \in C} \langle x^*, y \rangle .$$

Lemma 4.1

Assume that $F : E \to \mathbb{R} \cup \{+\infty\}$ is proper, convex, and lower semicontinuous. Let $x \in \text{dom}\,(F)$ be fixed. Then for any $d > 0$ there exists $x^* \in E^*$ such that

$$F(v) > F(x) + \langle x^*, v - x \rangle - d \text{ for all } v \in E. \qquad (4.1)$$

Proof Let $d > 0$. Observe that the set $\text{Epi}\,(F)$ is convex and closed and therefore, according to Lemma 2.1, sequentially weakly closed. We see that $(x, F(x) - d) \notin \text{Epi}\,(F)$. Therefore from Theorem 4.3 there exists a pair $(\bar{x}^*, a^*) \in E^* \times \mathbb{R}$ such that

$$\langle \bar{x}^*, x \rangle + a^*\,(F(x) - d) > \sup_{(y,a)\in\text{Epi}(F)} \left(\langle \bar{x}^*, y \rangle + a^* a \right).$$

By the above and since $(x, F(x)) \in \text{Epi}\,(F)$, we see that $a^*\,(-d) > 0$, i.e. $a^* < 0$. This means that for $x^* = -\frac{\bar{x}^*}{a^*}$ we have

$$\langle x^*, x \rangle - (F(x) - d) > \sup_{(y,a)\in\text{Epi}(F)} \left(\langle x^*, y \rangle - a \right) \geq$$
$$\langle x^*, v \rangle - F(v) \text{ for all } v \in E.$$

Hence (4.1) follows.

Remark 4.1

In case $x^* \in E^*$ is such that

$$F(v) \geq F(x) + \langle x^*, v - x \rangle \text{ for all } v \in E,$$

then it is called a supporting functional for functional $F : E \to \mathbb{R} \cup \{+\infty\}$ at point x.

Exercise 4.3 Prove that a proper convex and Gâteaux differentiable functional has exactly one supporting functional, i.e. its Gâteaux derivative. Hint: use definition of a first Gâteaux variation, see Definition 2.7.

Let us recall that a closed hyperplane $H \subset E \times \mathbb{R}$ has the following form:

$$H = \left\{ (x, t) \in E \times \mathbb{R} : \langle x^*, x \rangle + \beta = \alpha t \right\}$$

for some fixed $\alpha, \beta \in \mathbb{R}$ and a fixed $x^* \in E^*$.

If $\alpha \neq 0$, then the hyperplane is called nonvertical and has the form

$$H = \left\{ (x, t) \in E \times \mathbb{R} : \langle x^*, x \rangle + \beta = t \right\}$$

for some (possibly other than above) $\beta \in \mathbb{R}$ and $x^* \in E^*$.

Let $F : E \rightarrow \mathbb{R} \cup \{+\infty\}$. A nonvertical hyperplane H in $E \times \mathbb{R}$ is called supporting hyperplane for Epi (F) at the point $(x_0, F(x_0))$ if $(x_0, F(x_0)) \in H$ and if set Epi (F) is contained in one of the two closed subspaces determined by H. In this case H has the following form:

$$H = \left\{ (x, t) \in E \times \mathbb{R} : \langle x^*, x - x_0 \rangle + F(x_0) = t \right\}$$

and moreover

$$\text{Epi}\,(F) \subset \left\{ (x, t) \in E \times \mathbb{R} : \langle x^*, x - x_0 \rangle + F(x_0) \leq t \right\}.$$

4.2   On the Conjugate and Its Properties

Definition 4.1 (Fenchel-Young Conjugate)

The Fenchel-Young conjugate $F^* : E^* \rightarrow \mathbb{R} \cup \{+\infty\}$ for functional $F : E \rightarrow \mathbb{R} \cup \{+\infty\}$ reads as follows:

$$F^* \left(v^* \right) = \sup_{u \in E} \left\{ \langle v^*, u \rangle - F(u) \right\}. \tag{4.2}$$

When F is proper, the above definition can be given equivalently by taking supremum over dom (F). The formula (4.2) provides a sort of $1 - 1$ correspondence between convex sequentially weakly lower semicontinuous functionals considered on E and on E^*. When for a certain $v^* \in E^*$ the value $F^* (v^*)$ is attained at some u_0, then (4.2) means that

$$H = \left\{ (x, t) \in E \times \mathbb{R} : \langle v^*, x \rangle - F^* \left(v^* \right) = t \right\}$$

is a supporting hyperplane of Epi (F) at $(u_0, F(u_0))$.

Let $F, G : E \rightarrow \mathbb{R} \cup \{+\infty\}$. Directly from the definition of the Fenchel-Young conjugate we can obtain what follows:

(i) $\inf_{x \in E} F(x) = -F^* (0)$;
(ii) $F \leq G$ implies that $G^* \geq F^*$;
(iii) $(\lambda F)^* (x^*) = \lambda F^* (x^*/\lambda)$ for all $\lambda > 0$, $x^* \in E^*$;
(iv) $(F_h)^* (x^*) = F^* (x^*) + \langle x^*, h \rangle$ for all $x^* \in E^*$, $x, h \in E$ and where $F_h(x) = F(x - h)$.

The exercises that follow allow us to have some insight into the notion of the Fenchel-Young conjugate.

Exercise 4.4 Consider the function $F : \mathbb{R} \to \mathbb{R}$ defined by $F(x) = e^x$. Calculate that $F^* : \mathbb{R} \to \mathbb{R} \cup \{+\infty\}$ is given by the following formula:

$$
F^*(v) = \begin{cases} v \ln v - v, & v > 0, \\ 0, & v = 0, \\ +\infty, & \text{otherwise.} \end{cases}
$$

Exercise 4.5 Consider the function $F : \mathbb{R} \to \mathbb{R}$ defined by

$$
F(v) = \begin{cases} -\ln v, & v > 0, \\ +\infty, & \text{otherwise.} \end{cases}
\tag{4.3}
$$

Calculate F^*.

Exercise 4.6 Consider the function $G : \mathbb{R}^N \to \mathbb{R}$ defined by

$$
G(x) = G(x_1, x_2, \ldots, x_N) = \sum_{i=1}^N F(x_i),
$$

where F is given by (4.3). Calculate that $G^* : \mathbb{R}^N \to \mathbb{R}$ has the following form:

$$
G^*(v) = \sum_{i=1}^N F^*(v_i).
$$

Exercise 4.7 Let $F : \mathbb{R} \to \mathbb{R}$ be a convex and a continuous function. Derive the formula for the Fenchel-Young conjugate of $G : \mathbb{R}^N \to \mathbb{R}$ defined by $G(x) = \sum_{i=1}^N F(x_i)$.

The Fenchel-Young Inequality, which now follows despite being simple to be derived, is a very powerful tool.

Lemma 4.2 (Fenchel-Young Inequality)
Let $F : E \to \mathbb{R} \cup \{+\infty\}$ be any functional. Then for all $v \in E$, $v^* \in E^*$ we have

$$
F^*(v^*) + F(v) \geq \langle v^*, v \rangle.
$$

Proof We see from (4.2) that for any $v \in E$, $v^* \in E^*$

$$F(v) + F^*(v^*) = F(v) + \sup_{u \in E} \{\langle v^*, u \rangle - F(u)\} \geq$$
$$F(v) + \langle v^*, v \rangle - F(v) = \langle v^*, v \rangle.$$

There are striking special cases of the Fenchel-Young Inequality which are very well known and which can be obtained via other arguments as well.

Example 4.1 Let E be real Hilbert space identified with its dual via the Riesz Representation Theorem. Define $F : E \to \mathbb{R}$ by $F(x) = \frac{1}{2} \|x\|^2$. Then a direct calculation (using, for example, Theorem 2.22) reveals that $F^*(x^*) = \frac{1}{2} \|x^*\|^2$. Hence $F = F^*$. On the other hand if $F = F^*$ in the Hilbert space setting, then it can be proved that $F(x) = \frac{1}{2} \|x\|^2$. The Fenchel-Young Inequality now implies the obvious inequality

$$\|x^*\|^2 + \|x\|^2 \geq 2 \langle x^*, x \rangle \text{ for all } x \in E, \ x^* \in E.$$

For some more general version of this result see [28], where the so-called at most quadratic functions are considered.

Example 4.2 Consider a function $F : \mathbb{R} \to \mathbb{R}$ given by $F(x) = \frac{1}{p} |x|^p$ for some $p > 1$. By a direct calculation we see that $F^*(x^*) = \frac{1}{q} |x^*|^q$ and the Fenchel-Young Inequality now implies the well known inequality

$$ab \leq \frac{1}{p} a^p + \frac{1}{q} b^q \text{ for all } a, b > 0.$$

Having defined F^* we see that the definition of F^{**} is now obvious. Function

$$F^{**} : E \to \mathbb{R} \cup \{+\infty\}$$

can be viewed as a type of a closed convex envelope of a mapping F. However, the following exercise may be of interest in relating F and F^{**} in case F is not lower semicontinuous.

Remark 4.2

The Fenchel-Young conjugate has also other links to well know notions. Let us recall that for every $x \in E$ it holds

$$\|x\| = \sup_{l \in E^*, \|l\|_* \leq 1} |\langle l, x \rangle| = \max_{l \in E^*, \|l\|_* \leq 1} |\langle l, x \rangle|.$$

Let us consider $F : E \to \mathbb{R}$ given by $F(x) = \|x\|$. Then we calculate that $F^* : E^* \to \mathbb{R}$ is as follows:

(continued)

Remark 4.2 (continued)

$$F^*(x^*) = \begin{cases} 0, & \text{for } \|x^*\|_* \leq 1, \\ +\infty, & \text{for } \|x^*\|_* > 1. \end{cases}$$

Calculating F^{**} we see that

$$F^{**}(x) = \sup_{x^* \in E^*, \|x^*\|_* \leq 1} |\langle x^*, x \rangle| = \|x\|.$$

As expected from the geometric interpretation the following basic properties of the Fenchel-Young conjugate arise:

Lemma 4.3

Let $F : E \to \mathbb{R} \cup \{+\infty\}$ be proper. Then its Fenchel-Young conjugate $F^* : E^* \to \mathbb{R} \cup \{+\infty\}$ is convex and sequentially weakly lower semicontinuous. If additionally F is convex and lower semicontinuous, then F^* is proper as well.

Proof The convexity of F^* is apparent from the definition. Indeed, take $t \in (0, 1)$ and define $x_t^* = tx^* + (1 - t) y^*$ for arbitrary $x^*, y^* \in E^*$. Then we have

$$F^*(x_t^*) = \sup_{u \in E} \{\langle x_t^*, u \rangle - F(u)\} \leq$$
$$t \sup_{u \in E} \{\langle x^*, u \rangle - F(u)\} + (1 - t) \sup_{u \in E} \{\langle y^*, u \rangle - F(u)\} =$$
$$t F^*(x^*) + (1 - t) F^*(y^*).$$

In order to prove that F^* is sequentially weakly lower semicontinuous it suffices to demonstrate that $\text{Epi}(F^*)$ is sequentially weakly closed. Now we take a weakly convergent sequence $((x_n^*, a_n^*))_{n=1}^{\infty} \subset E^* \times \mathbb{R}$, where $x_n^* \rightharpoonup x_0^*$ and $a_n^* \to a_0^*$ are such that $F^*(x_n^*) \leq a_n^*$. We need to show that $(x_0^*, a_0^*) \in \text{Epi}(F^*)$, i.e. that $F^*(x_0^*) \leq a_0^*$. Suppose to the contrary that it holds $F^*(x_0^*) = a_0^* + \varepsilon$ for some $\varepsilon > 0$. Using the definition of F^* we see that there is some $x_0 \in \text{dom}(F)$ such that

$$\langle x^*, x_0 \rangle - F(x_0) \geq a_0^* + \frac{\varepsilon}{2}.$$

Then we have

$$F^*(x_n^*) = \sup_{u \in E} \{\langle x_n^*, u \rangle - F(u)\} \geq \langle x_n^*, x_0 \rangle - F(x_0) =$$
$$\langle x^*, x_0 \rangle - F(x_0) + \langle x_n^* - x^*, x_0 \rangle \geq a_0^* + \frac{\varepsilon}{2} + \langle x_n^* - x^*, x_0 \rangle.$$

Therefore we obtain

$$a_0^* = \lim_{n \to +\infty} a_n^* \geq \limsup_{n \to +\infty} F^*(x_n^*) \geq \lim_{n \to +\infty} \left(a_0^* + \frac{\varepsilon}{2} + \langle x_n^* - x^*, x_0 \rangle \right) \geq a_0^* + \frac{\varepsilon}{2}.$$

This contradiction proves that F^* is sequentially weakly lower semicontinuous.

Now we proceed with showing that F^* is proper under additional assumption that F is convex and lower semicontinuous. Since F is proper, there is some x_0 that $F(x_0) \in \mathbb{R}$. By Lemma 4.1 for any $d > 0$ there is $x^* \in E^*$ such that (4.1) holds, i.e.

$$\langle x^*, x_0 \rangle - F(x_0) + d > \langle x^*, v \rangle - F(v) \text{ for all } v \in E.$$

This means that

$$F^*(x^*) = \sup_{v \in E} \{\langle x^*, v \rangle - F(v)\} \leq \langle x^*, x_0 \rangle - F(x_0) + d$$

and therefore $F^*(x^*) \in \mathbb{R}$.

Remark 4.3
When F is not convex, then F^* need not be proper as seen for $F(x) = x^3$.

Exercise 4.8   Consider the function $F : \mathbb{R} \to \mathbb{R} \cup \{+\infty\}$ given by the following formula:

$$F(x) = \begin{cases} 0, & x < 0, \\ 1, & x = 0, \\ +\infty, & \text{otherwise.} \end{cases}$$

Show that the epigraph of F is not closed. Calculate the first and the second conjugate. Show that the epigraph of F^{**} is closed. Hint: some sketch of a graph may help in solving this problem.

Theorem 4.4   (Fenchel-Moreau Theorem)
Let $F : E \to \mathbb{R} \cup \{+\infty\}$ be proper. Then the following are equivalent:

(i) F is convex and sequentially weakly lower semicontinuous;
(ii) $F = F^{**}$.

Proof We start with proving that (i) implies (ii). We see that by the Fenchel-Young Inequality, Lemma 4.2, it follows that

$$F^{**}(x) = \sup_{u^* \in E^*} \{\langle u^*, x \rangle - F^*(u^*)\} \leq F(x).$$

Suppose F^{**} is finite at x and that $F^{**}(x) < F(x)$. Then

$$\left(x, F^{**}(x)\right) \notin \text{Epi}(F)$$

and since Epi (F) is convex and closed from Theorem 4.3 we see that there is a pair

$$\left(x^*, a^*\right) \in E^* \times \mathbb{R}$$

such that

$$\langle x^*, x\rangle + a^* F^{**}(x) > \sup_{(y,a)\in\text{Epi}(F)} \left(\langle x^*, y\rangle + a^* a\right).$$

As in the proof of Lemma 4.1 we may assume that $a^* = -1$. This shows that

$$\langle x^*, x\rangle - F^{**}(x) > \sup_{(y,a)\in\text{Epi}(F)} \left(\langle x^*, y\rangle - a\right) =$$
$$\sup_{y\in\text{dom}(F)} \left(\langle x^*, y\rangle - F(y)\right) = F^*(x^*).$$

Using the Fenchel-Young Inequality, Lemma 4.2, we see that

$$\langle x^*, x\rangle - F^{**}(x) > F^*\left(x^*\right) \geq -F^{**}(x) + \langle x^*, x\rangle$$

which is a contradiction. Therefore $F^{**}(x) \geq F(x)$ which provides that $F^{**}(x) = F(x)$.

Now we show that (ii) implies (i). Observe that functional F^{**} as a conjugate to F^* is sequentially weakly lower semicontinuous and convex.

Potential Operators

5

In this chapter we are interested in mappings that are derivatives of certain functionals. Given a functional defined on E, under suitable conditions, it is not a difficult task to investigate its differentiability. On the other hand, given a mapping acting between E and E^* it may be quite challenging to determine whether there exists a functional whose derivative such a mapping is. We will call such mappings potential. Following mainly [19], we present the theory of potential operators as well as comment on their invertibility.

5.1 Basic Concepts and Properties

The concepts which we introduce here allow for connecting monotonicity methods with variational ones.

Definition 5.1 (Potential Operator)
Operator $A : E \to E^*$ is potential if there exists a functional $F : E \to \mathbb{R}$ which is differentiable in the sense of Gâteaux on E and is such that $F' = A$. Functional F is called the potential of A.

Many operators which we have already encountered are in fact potential. Let us examine the following simple and most apparent examples.

Example 5.1 From Example 3.2 we know that operator $-\Delta : H_0^1(0,1) \to H^{-1}(0,1)$ defined by

$$\langle -\Delta u, v \rangle_{H^{-1}, H_0^1} = \int_0^1 \dot{u}(t)\, \dot{v}(t)\, dt \text{ for } u, v \in H_0^1(0,1)$$

© The Author(s), under exclusive license to Springer Nature Switzerland AG 2021
M. Galewski, *Basic Monotonicity Methods with Some Applications*,
Compact Textbooks in Mathematics, https://doi.org/10.1007/978-3-030-75308-5_5

is strongly monotone and continuous. Using calculations as in Example 2.5 we see that $-\Delta$ stands for the Fréchet derivative of the following C^1 functional acting on $H_0^1 (0, 1)$:

$$F(u) = \frac{1}{2} \int_0^1 |\dot{u}(t)|^2 \, dt = \frac{1}{2} \|u\|_{H_0^1}^2.$$

This means that $-\Delta$ is potential.

Proposition 5.1
The functional $F : L^p (0, 1) \to \mathbb{R}$ given by the formula

$$F(u) = \frac{1}{p} \int_0^1 |u(t)|^p \, dt = \frac{1}{p} \|u\|_{L^p}^p \tag{5.1}$$

is continuously differentiable and for any $u, v \in L^p (0, 1)$ it holds

$$\left\langle F'(u), v \right\rangle = \int_0^1 |u(t)|^{p-2} u(t) v(t) \, dt. \tag{5.2}$$

Proof Take any $u, v \in L^p (0, 1)$. For a.e. $t \in [0, 1]$ we easily calculate that

$$\lim_{s \to 0} \frac{|u(t) + sv(t)|^p - |u(t)|^p}{s} = |u(t)|^{p-2} u(t) v(t).$$

Let us now fix $0 < |s| < 1$. For a.e. $t \in [0, 1]$ we have by the Lagrange Mean Value Theorem that there exists some $\theta = \theta(t, s)$ such that

$$\frac{|u(t) + sv(t)|^p - |u(t)|^p}{s} = |u(t) + \theta v(t)|^{p-2} (u(t) + \theta v(t)) v(t).$$

The above estimation further implies that

$$\left| \frac{|u(t) + sv(t)|^p - |u(t)|^p}{s} \right| \le 2^{p-2} \left(|u(t)|^{p-1} |v(t)| + |v(t)|^p \right).$$

By a direct calculation using the Hölder Inequality we have that function

$$t \mapsto |u(t)|^{p-1} |v(t)| + |v(t)|^p$$

is integrable over $[0, 1]$. Now using the Lebesgue Dominated Convergence Theorem we see that (5.2) holds.

Now we prove that the derivative is continuous. Let g be the Nemytskii operator induced by function

$$u \mapsto |u|^{p-2} u.$$

We observe that if $u_n \to u_0$ in $L^p(0,1)$, then $g(u_n) \to g(u_0)$ in $L^q(0,1)$. Indeed, when $u \in L^p(0,1)$ we calculate directly that $g(u) \in L^q(0,1)$. The claim follows by the Krasnosel'skii Theorem on the continuity of the Nemytskii operator, Theorem 2.13.

From the above Proposition it is easy to verify all details in the following example.

Example 5.2 Let $p \geq 2$. We see that operator $-\Delta_p : W_0^{1,p}(0,1) \to W^{-1,q}(0,1)$ defined by

$$\langle -\Delta_p(u), v \rangle = \int_0^1 |\dot{u}(t)|^{p-2} \dot{u}(t) \dot{v}(t) \, dt \text{ for } u, v \in W_0^{1,p}(0,1)$$

is potential. Its potential $F : W_0^{1,p}(0,1) \to \mathbb{R}$ reads

$$F(u) = \frac{1}{p} \int_0^1 |\dot{u}(t)|^p \, dt = \frac{1}{p} \|u\|_{W_0^{1,p}}^p.$$

When taking into account Lemma 3.2 one may easily solve the following exercise (which together with Proposition 5.2 provides that the assertion in Example 5.2 holds).

Exercise 5.1 Assume that X is a real Banach space. Let $L : E \to X$ be such a linear operator that for $x \in E$ it holds

$$\|x\|_E = \|Lx\|_X.$$

Assume $A : X \to X^*$ is potential with potential F. Show that operator $T = L^* A L$, $T : E \to E^*$, is potential. Hint: consider on E functional $x \mapsto F(Lx)$ and use the **Chain Rule**, Theorem 2.15.

There is a direct formula which allows us to derive the potential in case some continuity assumption is added. This is to no surprise if one recalls the Newton-Leibniz Theorem known from the calculus courses.

Lemma 5.1

If $A : E \to E^*$ is potential with the potential denoted by F and demicontinuous, then for any $v \in E$ it holds

$$F(v) = F(0) + \int_0^1 \langle A(sv), v \rangle \, ds. \tag{5.3}$$

Proof Fix $v \in E$ and consider a differentiable function $g : [0, 1] \to \mathbb{R}$ given by the formula $g(t) = F(tv)$. Then for $t \in [0, 1]$ we have calculating directly from a definition of a derivative that

$$g'(t) = \lim_{h \to 0} \frac{F(tv + hv) - F(tv)}{h} = \langle A(tv), v \rangle.$$

Since A is demicontinuous it follows that g' is continuous on $[0, 1]$ and therefore it follows

$$F(v) - F(0) = g(1) - g(0) = \int_0^1 g'(t) \, dt = \int_0^1 \langle A(tv), v \rangle \, dt.$$

Exercise 5.2 Verify that (5.3) coincides with the formulas for potential obtained in Examples 5.1 and 5.2.

In order to give an example of a more advanced potential mapping we recall that in Theorem 3.3 we considered for $p \geq 2$ operator

$$A : L^p(0, 1) \to L^q(0, 1),$$

given by

$$\langle A(u), v \rangle = \int_0^1 \varphi \left(t, |u(t)|^{p-1} \right) |u(t)|^{p-2} u(t) v(t) \, dt \tag{5.4}$$

under assumption $\mathbf{A}\varphi$ about a Carathéodory function $\varphi : [0, 1] \times \mathbb{R}_+ \to \mathbb{R}$, that is: $|\varphi(t, x)| \leq M$ for a.e. $t \in [0, 1]$, all $x \in \mathbb{R}_+$ for some fixed constant $M > 0$.

Theorem 5.1

Assume that condition $A\varphi$ is satisfied. Operator A given by (5.4) is potential with the potential $F : L^p (0, 1) \to \mathbb{R}$ defined by

$$F(u) = \int_0^1 \int_0^{|u(t)|} \varphi\left(t, s^{p-1}\right) s^{p-1} ds\, dt \text{ for } u \in L^p (0, 1).$$

Proof For any fixed $u, v \in L^p (0, 1)$ we have by a direct differentiation

$$\lim_{\lambda \to 0} \frac{F(u+\lambda v) - F(u)}{\lambda} = \frac{d}{d\lambda} \int_0^1 \int_0^{|u(t)+\lambda v(t)|} \varphi\left(x, s^{p-1}\right) s^{p-1} ds\, dt \Big|_{\lambda=0} =$$
$$\int_0^1 \varphi\left(t, |u(t)|^{p-1}\right) |u(t)|^{p-2} u(t) v(t)\, dt = \langle A(u), v \rangle.$$

Observe that operator A being potential means the Gâteaux differentiability of F. If A is continuous, then F is a C^1 functional. Since the Gâteaux differentiability of a functional does not imply any type of its continuity, care must be taken into consideration of potential mappings. Note that the result which we provide now is a sort of converse of what we already know about the monotonicity of a derivative of a convex functional.

Lemma 5.2

Assume that $A : E \to E^*$ is potential and monotone. Then its potential $F : E \to \mathbb{R}$ is convex and sequentially weakly lower semicontinuous.

Proof The assumptions mean that the potential $F : E \to \mathbb{R}$ of operator A is differentiable in the sense of Gâteaux and convex. Let $(u_n)_{n=1}^\infty$ be weakly convergent to u_0. Then we have by Theorem 3.1 that for all $n \in \mathbb{N}$ it holds

$$F(u_n) \geq F(u_0) + \langle A(u_0), u_n - u_0 \rangle.$$

Passing to a limit we obtain that

$$\liminf_{n \to +\infty} F(u_n) \geq F(u_0) + \liminf_{n \to +\infty} \langle A(u_0), u_n - u_0 \rangle = F(u_0).$$

This proves the sequential weak lower semicontinuity of F.

Now we connect the notion of the potential of an operator with the solvability of a nonlinear equation which involves this operator. The results which follow are not unexpected but will be crucial when developing the theory further on.

Lemma 5.3

Assume that $A : E \to E^*$ is potential with potential F and let $g \in E^*$ be fixed. If u_0 is a minimizer over E of a functional

$$u \mapsto F(u) - \langle g, u \rangle,$$

i.e.

$$F(u_0) - \langle g, u_0 \rangle = \inf_{u \in E} \left(F(u) - \langle g, u \rangle \right), \tag{5.5}$$

then it follows that u_0 solves equation

$$A(u) = g. \tag{5.6}$$

On the other hand, if A is additionally monotone, and if u_0 solves (5.6), then it is also a minimizer to $u \mapsto F(u) - \langle g, u \rangle$.

Proof Assume that u_0 satisfies (5.5). Then for any $h \in E$ we see by the **Fermat Rule** that

$$0 = \lim_{t \to 0} \frac{F(u_0 + th) - \langle g, u_0 + th \rangle}{t} = \langle A(u_0) - g, h \rangle.$$

Therefore u_0 solves (5.6).

Now assume that A is monotone and that u_0 solves (5.6) which means that $A(u_0) = g$. From Lemma 5.2 we know that F is convex which means by Theorem 3.1 that for any $u \in E$

$$0 \le F(u) - F(u_0) - \langle A(u_0), u - u_0 \rangle =$$
$$F(u) - \langle g, u \rangle - F(u_0) - \langle g, u_0 \rangle$$

which is the second assertion.

We know that monotone operators may lack some suitable type of continuity, see Example 1.1. This, however, may not happen for monotone and potential operators.

Lemma 5.4

Assume that $A : E \to E^*$ is potential and monotone. Then A is demicontinuous.

Proof Using Lemma 3.6 it suffices to demonstrate that relation

$$\langle g - A(v), u - v \rangle \geq 0 \text{ for any } v \in E \qquad (5.7)$$

implies that $A(u) = g$, where $u \in E$, $g \in E^*$. Let us take in (5.7) $v_t = u + t(v - u)$ for $t > 0$ instead of v. Then we have

$$t \langle g - A(v_t), v - u \rangle \leq 0$$

which implies

$$\langle g, v - u \rangle \leq \langle A(v_t), v - u \rangle.$$

Since A is potential and monotone, its potential is convex and differentiable in the sense of Gâteaux. Thus we have from the above that

$$F(v_t + v - u) - F(v_t) \geq \langle A(v_t), v - u \rangle \geq \langle g, v - u \rangle.$$

Taking a limit as $t \to 0$ we obtain

$$F(v) - F(u) \geq \langle g, v - u \rangle.$$

Since v is arbitrary this means that $g = A(u)$ by Lemma 5.3.

Remark 5.1

The above result says that a potential and monotone mapping is necessarily demicontinuous. Since in a finite dimensional case demicontinuity is equivalent to continuity, we see at once that the operator given in Example 1.1 is not potential.

Now we turn to the coercivity of the potential. Note that the potential of a monotone operator need not be coercive, see, for example, the exponential function. The following exercises can be solved using our previous considerations.

Exercise 5.3 Assume that $A : E \to E^*$ is potential, coercive, and monotone. Show that its potential is bounded from below.

Exercise 5.4 Assume that $A : E \to E^*$ is potential and d–monotone, $A(0) = 0$ and $\rho(0) = 0$ in (3.1). Prove that its potential $F : E \to \mathbb{R}$ is coercive.

Exercise 5.5 Assume that $A : E \to E^*$ is potential and uniformly monotone. Prove that the potential of A is coercive.

With the above preparations we can proceed to some general case:

> **Lemma 5.5**
> Assume that $A : E \to E^*$ is potential, demicontinuous, bounded, and coercive. Then its potential $F : E \to \mathbb{R}$ is coercive.

Proof By Lemma 5.1 we obtain for $v \neq 0$ by the change of variables formula

$$F(v) - F(0) = \int_0^1 \langle A(sv), v \rangle \, ds = \int_0^{\|v\|} \left\langle A\left(t \frac{v}{\|v\|}\right), t \frac{v}{\|v\|} \right\rangle \frac{dt}{t}. \tag{5.8}$$

Since A is coercive, there is some $t_0 > 0$ that

$$\left\langle A\left(t \frac{v}{\|v\|}\right), t \frac{v}{\|v\|} \right\rangle \geq \left\| t \frac{v}{\|v\|} \right\|$$

for all $t > t_0$. Since operator A is bounded, we see that for $t \in [0, t_0]$ it holds

$$\left| \left\langle A\left(t \frac{v}{\|v\|}\right), \frac{v}{\|v\|} \right\rangle \right| \leq m$$

for some constant m (independent of v and t). Therefore from (5.8) we have what follows

$$F(v) - F(0) \geq \int_0^{t_0} \left\langle A\left(t \frac{v}{\|v\|}\right), t \frac{v}{\|v\|} \right\rangle \frac{dt}{t} + \int_{t_0}^{\|v\|} \left\langle A\left(t \frac{v}{\|v\|}\right), t \frac{v}{\|v\|} \right\rangle \frac{dt}{t} \geq$$
$$\|v\| - t_0 m - t_0.$$

The last inequality implies that F is coercive.

5.2   Invertible Potential Operators

Now we consider inversion of potential operators. This is done with the help of the Fenchel-Young conjugacy. The theorem which follows should be compared with the Fenchel-Moreau Theorem.

Theorem 5.2

Assume that $F : E \to \mathbb{R}$ is convex and sequentially weakly lower semicontinuous. Suppose that operator $A : E \to E^*$ is invertible with inverse $A^{-1} : E^* \to E$. Then the following are equivalent:

 (i) *A is a derivative of F;*
 (ii) *A is radially continuous and for all $x, y \in E$ it holds*

$$F(x) + F^*(A(x)) = \langle A(x), x \rangle,$$
$$F(y) \geq F(x) + \langle A(x), y - x \rangle;$$

(iii) *A^{-1} is radially continuous and for all $x^*, y^* \in E^*$ it holds*

$$F(A^{-1}(x^*)) + F^*(x^*) = \langle x^*, A^{-1}(x^*) \rangle,$$
$$F^*(y^*) \geq F^*(x^*) + \langle y^* - x^*, A^{-1}(x^*) \rangle;$$

(iv) *A^{-1} is a derivative of F^*.*

Proof Note that since F is convex, then its derivative, when exists, is a monotone operator.

We show first that (i) implies (ii). By Lemma 5.4 it follows that A is demicontinuous and therefore radially continuous. The second assertion in (ii) follows by Theorem 3.1. We prove the first assertion. By the Fenchel-Young Inequality, see Lemma 4.2, we obtain

$$F(x) + F^*(A(x)) \geq \langle A(x), x \rangle \text{ for all } x \in E.$$

By the convexity of F we see that for all $x, u \in E$

$$\langle A(x), x \rangle - F(x) \geq \langle A(x), u \rangle - F(u).$$

Using definition of F^* we obtain from the above

$$F(x) + F^*(A(x)) = F(x) + \sup_{u \in E} \{ \langle A(x), u \rangle - F(u) \} \leq \langle A(x), x \rangle.$$

Next we show that (ii) implies (iii). We note that for all $x, y \in E$ it holds

$$F(y) \geq F(x) + \langle A(x), y - x \rangle \text{ and } F(x) \geq F(y) + \langle A(y), x - y \rangle.$$

Combining the above inequalities we see that both A and A^{-1} are monotone due to Theorem 3.1. We will show that A^{-1} is radially continuous by applying the fundamental lemma, Lemma 3.6. Precisely, we will show that relation

$$\left\langle x^* - y^*, x - A^{-1}\left(y^*\right)\right\rangle \geq 0 \text{ for all } y^* \in E^*$$

implies that $A^{-1}(x^*) = x$ or else, equivalently, that $A(x) = x^*$. Putting $A(y) = y^*$ we obtain the equivalent relation

$$\left\langle x^* - A(y), x - y\right\rangle \geq 0 \text{ for all } y \in E$$

which implies that $A(x) = x^*$ since A is monotone and radially continuous. Putting $A^{-1}(x^*) = x$ in relation

$$F(x) + F^*(A(x)) = \langle A(x), x \rangle$$

we obtain

$$F\left(A^{-1}\left(x^*\right)\right) + F^*\left(x^*\right) = \left\langle x^*, A^{-1}\left(x^*\right)\right\rangle \text{ for all } x^* \in E^*.$$

Now we prove the second assertion in (iii). Let us take any $x^*, y^* \in E^*$ and define

$$A^{-1}\left(x^*\right) = x, \quad A^{-1}\left(y^*\right) = y.$$

Then it follows

$$F^*(y^*) - F^*(x^*) = -F(y) + F(x) + \langle A(y), y\rangle - \langle A(x), x\rangle =$$
$$-F(y) + F(x) - \langle A(y), x - y\rangle + \langle A(y) - A(x), x\rangle \geq$$
$$\langle A(y) - A(x), x\rangle = \left\langle y^* - x^*, A^{-1}\left(x^*\right)\right\rangle.$$

We proceed to showing that (iii) implies (iv). From (iii) we have the following inequalities:

$$F^*\left(y^*\right) - F^*\left(x^*\right) \geq \left\langle y^* - x^*, A^{-1}\left(x^*\right)\right\rangle$$

and

$$F^*\left(x^*\right) - F^*\left(y^*\right) \geq \left\langle x^* - y^*, A^{-1}\left(y^*\right)\right\rangle$$

which imply that

$$\left\langle y^* - x^*, A^{-1}\left(x^*\right)\right\rangle \leq F^*\left(y^*\right) - F^*\left(x^*\right) \leq \left\langle y^* - x^*, A^{-1}\left(y^*\right)\right\rangle.$$

We define $y^* = x^* + tz^*$, where $t > 0$, $z^* \in E^*$. Then we have

$$\left\langle z^*, A^{-1}\left(x^*\right)\right\rangle \leq \frac{F^*\left(x^* + tz^*\right) - F^*\left(x^*\right)}{t} \leq \left\langle z^*, A^{-1}\left(x^* + tz^*\right)\right\rangle.$$

Taking a limit as $t \to 0$ and using the fact that A^{-1} is radially continuous, we see that F^* is differentiable in the sense of Gâteaux with A^{-1} being its derivative.

Finally we show that (iv) implies (i). From the previous parts of the proof it follows that $A = \left(A^{-1}\right)^{-1}$ is the derivative of F^{**}. By Theorem 4.4 we see that $F = F^{**}$ and therefore we have the last assertion.

5.3   Criteria for Checking the Potentiality

We will introduce some sufficient conditions for potentiality of mappings. These are directly related to the well known notions of the path independence property of a conservative vector field to which these should be compared.

Lemma 5.6

Assume that $A : E \to E^*$ is radially continuous. Then the following are equivalent:

(i) operator A is potential;
(ii) for any $x, y \in E$ it holds

$$\int_0^1 \langle A(sx), x \rangle \, ds - \int_0^1 \langle A(sy), y \rangle \, ds = \int_0^1 \langle A(y + s(x - y)), x - y \rangle \, ds.$$

$$(5.9)$$

Proof We first show that (i) implies (ii). Denote by F the potential of A. Indeed, for any $x, y \in E$ it holds

$$\int_0^1 \langle A(sx), x \rangle \, ds - \int_0^1 \langle A(sy), y \rangle \, ds = F(x) - F(y) =$$
$$\int_0^1 \frac{d}{dt} F(y + s(x - y)) \, ds = \int_0^1 \langle A(y + s(x - y)), x - y \rangle \, ds.$$

Now we demonstrate that (ii) implies (i). We will show that functional F defined by

$$x \mapsto \int_0^1 \langle A(sx), x \rangle \, ds$$

is the potential of A. Let $x, y \in E$ be fixed. Using (5.9) and next using the integral mean value theorem we have for some suitable $s_0 \in [0, 1]$:

$$\lim_{t \to 0} \frac{F(x+ty)-F(x)}{t} = \lim_{t \to 0} \frac{1}{t} \int_0^1 \langle A(x+tsy), ty \rangle \, ds =$$
$$\lim_{t \to 0} \langle A(x+ts_0y), y \rangle = \langle A(x), y \rangle.$$

Thus the assertion is proved.

Let $C^1([0, 1], E)$ be a space of continuously differentiable mappings with values in E which equipped with a standard maximum norm

$$\|u\|_{C^1} := \max_{t \in [0,1]} \|u(t)\| + \max_{t \in [0,1]} \|\dot{u}(t)\|$$

becomes a Banach space.

Lemma 5.7
Assume that $A : E \to E^*$ is demicontinuous. Then conditions (i) and (ii) from Lemma 5.6 are equivalent to the following:

(iii) for all $x, y \in E$ and any $u \in C^1([0, 1], E)$ such that $u(0) = x$ and $u(1) = y$ it holds

$$\int_0^1 \langle A(sx), x \rangle \, ds - \int_0^1 \langle A(sy), y \rangle \, ds = \int_0^1 \langle A(u(s)), \dot{u}(s) \rangle \, ds.$$

Proof We will prove that (ii) implies (iii). Since A is demicontinuous we find a constant $M > 0$ such that for any $\tau, t, s \in [0, 1]$

$$\|A(u(s) + \tau(u(t) - u(s)))\|_* \le M.$$

Let us consider $\varphi : [0, 1] \to \mathbb{R}$ defined by

$$\varphi(t) = \int_0^1 \langle A(\tau u(t)), u(t) \rangle \, d\tau.$$

We see that from Lemma 5.6(ii) it follows for any $t, s \in [0, 1]$

$$|\varphi(t) - \varphi(s)| = \left| \int_0^1 \langle A(\tau u(t)), u(t) \rangle \, d\tau - \int_0^1 \langle A(\tau u(s)), u(s) \rangle \, d\tau \right| =$$
$$\left| \int_0^1 \langle A(u(s) + \tau(u(t) - u(s))), u(t) - u(s) \rangle \, d\tau \right| \le$$
$$M \|u(t) - u(s)\| \le M \max_{\tau \in [0,1]} \|\dot{u}(\tau)\| \, |t - s| \le M \|u\|_{C^1} \, |t - s|.$$

The above means φ is Lipschitz-continuous and therefore absolutely continuous which provides that φ has a derivative a.e. which is integrable over $[0, 1]$ that is

$$\int_0^1 \langle A(\tau x), x \rangle \, d\tau - \int_0^1 \langle A(\tau y), y \rangle \, d\tau = \varphi(1) - \varphi(0) = \int_0^1 \dot{\varphi}(t) \, dt. \qquad (5.10)$$

Moreover, we have for some suitable $\tau_0 \in [0, 1]$ from the Integral Mean Value Theorem

$$\lim_{s \to t} \frac{\varphi(s) - \varphi(t)}{s - t} = \lim_{s \to t} \frac{\int_0^1 \langle A(u(t) + \tau(u(s) - u(t))), u(s) - u(t) \rangle d\tau}{s - t} =$$
$$\lim_{s \to t} \left\langle A(u(t) + \tau_0(u(s) - u(t))), \frac{u(s) - u(t)}{s - t} \right\rangle =$$
$$\langle A(u(t)), \dot{u}(t) \rangle.$$

This means that

$$\dot{\varphi}(t) = \lim_{s \to t} \frac{\varphi(s) - \varphi(t)}{s - t} = \langle A(u(s)), \dot{u}(t) \rangle.$$

Thus from (5.10) we have the assertion.

It is obvious to observe that (iii) implies (ii). $\qquad\square$

There is also a sufficient condition which involves differentiability of A. The type of continuity which we impose on the derivative of A is to be compared with Proposition 3.2.

Lemma 5.8

Assume that $A : E \to E^*$ is a Gâteaux differentiable operator such that for any $x, y, h \in E$ function

$$(s, t) \mapsto \langle A'(h + sx + ty) x, y \rangle$$

is continuous on $[0, 1] \times [0, 1]$. Then the following are equivalent:

(i) operator A is potential;
(ii) for any $x, y, h \in E$ it holds

$$\langle A'(h) x, y \rangle = \langle A'(h) y, x \rangle.$$

Proof We start with proving that (i) implies (ii). Denote by F the potential of A. We define on $[0, 1]$ for fixed $x, y \in E$

$$\varphi_{h,x,y}(t) = F(h + tx + ty) - F(h + tx) - (F(h + ty) - F(h))$$

and use formula (5.9) in order to obtain

$$\varphi_{h,x,y}(t) = \int_0^1 \langle A(h+tx+sty), ty \rangle \, ds - \int_0^1 \langle A(h+sty), ty \rangle \, ds =$$
$$\int_0^t \langle A(h+tx+s_1 y) - A(h+s_1 y), y \rangle \, ds_1 =$$
$$\int_0^t \int_0^t \langle A'(h+s_1 y+s_2 x)x, y \rangle \, ds_2 ds_1 = t^2 \langle A'(h+\tau_1 y+\tau_2 x)x, y \rangle,$$

where the last equality is obtained by the Integral Mean Value Theorem for some suitable $\tau_1, \tau_2 \in [0, t]$. We see that

$$\varphi_{h,x,y} = \varphi_{h,y,x}$$

which results in the following equality:

$$t^2 \langle A'(h+\tau_1 y+\tau_2 x)x, y \rangle = t^2 \langle A'(h+\tau_3 y+\tau_4 x)y, x \rangle$$

for some suitable $\tau_3, \tau_4 \in [0, t]$. Then obviously

$$\langle A'(h+\tau_1 y+\tau_2 x)x, y \rangle = \langle A'(h+\tau_3 y+\tau_4 x)y, x \rangle$$

and letting $t \to 0^+$ we see that

$$\langle A'(h)x, y \rangle = \langle A'(h)y, x \rangle.$$

Now we prove that (ii) implies (i) using Lemma 5.6. For any $x, y \in E$ we have

$$\int_0^1 \langle A(tx), x \rangle \, dt - \int_0^1 \langle A(ty), y \rangle \, dt =$$
$$\int_0^1 \langle A(tx), x-y \rangle \, dt + \int_0^1 \langle A(tx) - A(ty), y \rangle \, dt =$$
$$\int_0^1 \langle A(tx), x-y \rangle \, dt + \int_0^1 \int_0^t \langle A'(ty+s(x-y))(x-y), y \rangle \, ds dt =$$
$$\int_0^1 \langle A(tx), x-y \rangle \, dt + \int_0^1 \int_s^1 \langle A'(ty+s(x-y))y, x-y \rangle \, dt ds =$$
$$\int_0^1 \langle A(sx), x-y \rangle \, ds + \int_0^1 \langle A(y+s(x-y)) - A(sx), x-y \rangle \, ds =$$
$$\int_0^1 \langle A(y+s(x-y)), x-y \rangle \, ds.$$

Then we get the assertion by (5.9).

Exercise 5.6 Let $p \geq 2$. Calculate the second variation of (5.1). Using the formula obtained and Lemma 5.8 prove that operator $-\Delta_p : W_0^{1,p}(0, 1) \to W^{-1,q}(0, 1)$ is potential.

Exercise 5.7 Assume that $\varphi : [0, 1] \times \mathbb{R}_+ \to \mathbb{R}$ is a Carathéodory function such that $x \mapsto \varphi(t, x)$ is continuously differentiable for $a.e.$ $t \in [0, 1]$. Assume also that there exists a constant $M > 0$ such that for a.e. $t \in [0, 1]$ and all $x \in \mathbb{R}_+$ it holds

$$|\varphi(t, x)| \leq M \text{ and } \left| \frac{\partial}{\partial x} \varphi(t, x) x \right| \leq M$$

Using Lemma 5.8 and Theorem 3.3(vi) prove that operator

$$A : W_0^{1,p} (0, 1) \to W^{-1,q} (0, 1)$$

defined for any $u, v \in W_0^{1,p} (0, 1)$ by

$$\langle A (u), v \rangle = \int_0^1 \varphi \left(t, |\dot{u} (t)|^{p-1} \right) |\dot{u} (t)|^{p-2} \dot{u} (t) \dot{v} (t) \, dt$$

is potential.

Existence of Solutions to Abstract Equations

6

We have already obtained the existence result involving equations containing continuous, monotone, and coercive mappings in finite dimensional spaces. Now, we are interested in shifting this result to the infinite space. Starting from the simplest existence theorem which follows directly from the Banach Contraction Principle and which comes from [57], we introduce several theorems concerning the existence of solutions to nonlinear equations. This chapter is written following [14, 17, 19]. The approach toward Leray–Lions Theorem is taken after [50].

As we shall see below, the main idea lying behind the existence of solutions to nonlinear equations via monotonicity theory is the fact that a solution is approximated by some, at least weakly, convergent sequence.

6.1 Preliminary Result

Our first existence result is a simple one. For its proof, we will need the following version of the Banach Contraction Principle (see Theorem 2.3.1 in [14]):

Theorem 6.1 (Banach Contraction Principle)
Let E be a Banach space. Assume that $A : E \to E$ is a contraction. Then mapping A has a unique fixed point, i.e. there is $x_0 \in E$ such that

$$A(x_0) = x_0.$$

© The Author(s), under exclusive license to Springer Nature Switzerland AG 2021
M. Galewski, *Basic Monotonicity Methods with Some Applications*,
Compact Textbooks in Mathematics, https://doi.org/10.1007/978-3-030-75308-5_6

Theorem 6.2

Assume E is a Hilbert space and $A : E \to E$ is Lipschitz continuous, i.e. there is $M > 0$ such that for all $u, v \in E$, we have

$$\| A(u) - A(v) \| \leq M \| u - v \|$$

and strongly monotone with a constant $m < M$, i.e. for $u, v \in E$, we have

$$(A(u) - A(v), u - v)_E \geq m \| u - v \|^2.$$

Then, for each $h \in E$, equation

$$A(u) = h \tag{6.1}$$

has exactly one solution. Moreover, A is invertible and $A^{-1} : E \to E$ is Lipschitz continuous.

Proof Let

$$0 < \varepsilon < \frac{2m}{M^2}, \tag{6.2}$$

and let us define an operator $T_\varepsilon : E \to E$ by

$$T_\varepsilon(u) = u - \varepsilon(A(u) - h).$$

We see that for all $u, v \in E$,

$$\| T_\varepsilon(u) - T_\varepsilon(v) \|^2 =$$
$$\| u - v \|^2 + \varepsilon^2 \| A(u) - A(v) \|^2 - 2\varepsilon(A(u) - A(v), u - v)_E \leq$$
$$\left(1 + M^2 \varepsilon^2 - 2m\varepsilon \right) \| u - v \|^2.$$

By (6.2), it follows that $1 + \varepsilon^2 M^2 - 2m\varepsilon < 1$. Therefore by the Banach Contraction Principle, there is exactly one u_ε such that $u_\varepsilon = T_\varepsilon(u_\varepsilon)$, which means that

$$u_\varepsilon = u_\varepsilon - \varepsilon(A(u_\varepsilon) - h).$$

This provides the unique solvability of (6.1).

The Lipschitz continuity of A^{-1} is demonstrated as follows. Let us observe that for any $u, v \in E$ from relation,

$$m \| u - v \|^2 \leq (A(u) - A(v), u - v)_E \leq \| A(u) - A(v) \| \, \| u - v \|,$$

we have

$$\|A(u) - A(v)\| \geq m \|u - v\|.$$

Putting $u = A^{-1}(x)$ and $v = A^{-1}(y)$, we see

$$\left\| A^{-1}(x) - A^{-1}(y) \right\| \leq \frac{1}{m} \|x - y\|.$$

Hence the Lipschitz continuity of A^{-1} follows.

Remark 6.1

From the proof of the Banach Contraction Principle, it follows that for each $u_0 \in E$ and each ε satisfying (6.2), the iteration method

$$u_{n+1} = u_n - \varepsilon(A(u_n) - h) \text{ for each } n = 0, 1, 2, \ldots$$

converges to the unique solution u of (6.1) with the a priori error estimate

$$\|u - u_n\| \leq \varepsilon^n (1 - \varepsilon)^{-1} \|u_1 - u_0\|$$

and a posteriori error estimate

$$\|u - u_{n+1}\| \leq \varepsilon(1 - \varepsilon)^{-1} \|u_{n+1} - u_n\|.$$

The rate of convergence is as follows:

$$\|u - u_{n+1}\| \leq \varepsilon \|u - u_n\|.$$

6.2 The Browder–Minty Theorem

In this section, we will give the main existence tool proved via finite dimensional approximations that give some clue how to define the so-called Galerkin type approximations. Some technical lemmas are in order for the proof of the main existence result.

Lemma 6.1

Assume that $A : E \to E^*$ is coercive. Then, for any $f \in E^*$, the set K of solutions to $A(u) = f$ is bounded.

Proof Assume that u is a solution to $A(u) = f$. Then, obviously

$$\langle A(u), u \rangle \leq \|f\|_* \|u\|. \tag{6.3}$$

Suppose that K is unbounded, i.e. there is a sequence $(u_n)_{n=1}^{\infty}$ such that $\|u_n\| \to +\infty$. Since A is coercive, there is some $N_0 > 0$ such that for all $n \geq N_0$, it holds that

$$\langle A(u_n), u_n \rangle \geq 2\|f\|_* \|u_n\|,$$

and we have contradiction with (6.3).

Exercise 6.1 Check if the assertion of the above lemma holds under assumption that $A : E \to E^*$ is weakly coercive.

We have considered the finite dimensional version of the below lemma with a different proof. What we show now is a sort of a direct proof using the developed tools for monotone operators.

> **Lemma 6.2**
>
> *Assume that $A : E \to E^*$ is radially continuous and monotone. Then, for any $f \in E^*$, the set K of solutions to $A(u) = f$ is sequentially weakly closed and convex.*

Proof Let $u_1, u_2 \in K$, i.e. $A(u_i) = f$ for $i = 1, 2$. Take $t \in (0, 1)$ and consider

$$u_t = tu_1 + (1 - t)u_2.$$

Observe that

$$f = tA(u_1) + (1 - t)A(u_2).$$

We see that any $v \in E$

$$\langle f - A(v), tu_1 + (1 - t)u_2 - v \rangle = $$
$$t\langle A(u_1) - A(v), u_1 - v \rangle + (1 - t)\langle A(u_2) - A(v), u_2 - v \rangle.$$

By the monotonicity, we obtain

$$\langle f - A(v), tu_1 + (1 - t)u_2 - v \rangle \geq 0,$$

and since A is radially continuous, it follows by Lemma 3.6, condition (ii), that $A(u_t) = f$.

Now take a weakly convergent sequence $(u_n)_{n=1}^{\infty} \subset K$, i.e. $A(u_n) = f$ for $n = 1, 2, \ldots$ Let u_0 denote its limit. Then, we have

$$\langle f - A(v), u_0 - v \rangle = \lim_{n \to +\infty} \langle f - A(v), u_n - v \rangle =$$
$$\lim_{n \to +\infty} \langle A(u_n) - A(v), u_n - v \rangle \geq 0.$$

By Lemma 3.6, condition (ii), we see that $A(u_0) = f$.

We will also need the following technical result:

Lemma 6.3
Assume that operator $A : E \to E^*$ is monotone and $K \subset E$ is such a set that

$$\|u\| \leq M_1 \text{ and } \langle A(u), u \rangle \leq M_2 \text{ for each } u \in K,$$

where M_1 and M_2 are some constants. Then there exists a constant M such that

$$\|A(u)\|_* \leq M \text{ for all } u \in K.$$

Proof Since by Proposition 3.3 operator A is locally bounded, there are constants $\varepsilon > 0$ and $M_3 > 0$ such that

$$\|A(u)\|_* \leq M_3$$

provided that $\|u\| \leq \varepsilon$. Using the monotonicity of A, we see that for any $u \in K$,

$$\|A(u)\|_* = \sup_{\|y\| \leq \varepsilon} \frac{1}{\varepsilon} \langle A(u), y \rangle \leq$$
$$\sup_{\|y\| \leq \varepsilon} \frac{1}{\varepsilon} (\langle A(u), u \rangle + \langle A(y), y \rangle - \langle A(y), u \rangle) \leq$$
$$\frac{1}{\varepsilon} (M_2 + \varepsilon M_3 + M_1 M_3) := M.$$

This finishes the proof.

Remark 6.2
Since E is separable, it contains a dense and countable set $\{h_1, \ldots, h_n, \ldots\}$. Define E_n for $n \in \mathbb{N}$ as a linear hull of $\{h_1, \ldots, h_n\}$. The sequence of subspaces E_n has the approximation property: for each $u \in E$, there is a sequence $(u_n)_{n=1}^{\infty}$ such that $u_n \in E_n$ for $n \in \mathbb{N}$ and $u_n \to u$.

We will employ these observations in the sequel. Now we can proceed with the formulation and the proof of the famous Browder–Minty Theorem. We will use the already established 1.2, which concerned the finite dimensional case.

Theorem 6.3 (Browder–Minty Theorem)
Assume that $A : E \to E^*$ is radially continuous, coercive, and monotone. Then for any $f \in E^*$, the set K of solutions to $A(u) = f$ is non-empty, bounded, sequentially weakly closed, and convex.

Proof By Lemmas 6.1 and 6.2, it follows that we need to examine solvability of the equation $A(u) = f$. Let us fix $n \in \mathbb{N}$ and space E_n from Remark 6.2. By f_n, we denote the restriction of functional f to space E_n. Similarly by A_n, we understand the restriction of A to space E_n. Then

$$A_n : E_n \to E_n^*.$$

Note that since A is coercive and monotone, so is A_n. Moreover, since by Lemma 3.6 A is also demicontinuous, it follows that A_n is continuous. Then by Theorem 1.2 we see that equation

$$A_n(u) = f_n \tag{6.4}$$

has at least one solution u_n. By definition of A_n, this means that

$$\langle A(u_n), h_k \rangle = \langle f, h_k \rangle \text{ for } k = 1, 2, \dots, n. \tag{6.5}$$

We therefore obtain a sequence $(u_n)_{n=1}^{\infty}$ of solutions to (6.4). We see that

$$\langle A(u_n), u_n \rangle = \langle f_n, u_n \rangle = \langle f, u_n \rangle \leq \|f\|_* \|u_n\|. \tag{6.6}$$

Since operator A is coercive, we observe by Lemma 6.1 that sequence $(u_n)_{n=1}^{\infty}$ is bounded. Thus there is some $R > 0$ that

$$\|u_n\| \leq R \text{ for } n \in \mathbb{N}.$$

From (6.6), it follows that

$$\langle A(u_n), u_n \rangle = \langle f, u_n \rangle \leq \|f\|_* R.$$

Since $\|u_n\| \le R$ and since

$$\langle A(u_n), u_n \rangle \le \|f\|_* R,$$

it follows by Lemma 6.3 that $\|A(u_n)\|_* \le M_1$ for some fixed $M_1 > 0$ and for all $n \in \mathbb{N}$. Next, we see that from (6.5), it follows

$$\lim_{n \to +\infty} \langle A(u_n), h \rangle = \langle f, h \rangle \quad \text{for each } h \in \bigcup_n E_n.$$

This means that $A(u_n) \rightharpoonup f$. Since $(u_n)_{n=1}^{\infty}$ is bounded, there is a subsequence $(u_{n_k})_{k=1}^{\infty}$ convergent weakly to some u_0. Again from (6.5), we see for all $k \in \mathbb{N}$ that

$$\langle A(u_{n_k}), u_{n_k} \rangle = \langle f, u_{n_k} \rangle.$$

Passing to a limit, we obtain

$$\lim_{k \to +\infty} \langle A(u_{n_k}), u_{n_k} \rangle = \langle f, u_0 \rangle$$

Hence, by Lemma 3.6(iii), we observe that $A(u_0) = f$.

It is easy to obtain the uniqueness in the above Browder–Minty Theorem as a consequence of strict monotonicity.

Corollary 6.1

Assume that $A : E \to E^*$ is radially continuous, coercive, and strictly monotone. Then for any $f \in E^*$, equation

$$A(u) = f \tag{6.7}$$

has exactly one solution.

Proof Suppose both $u, v \in E, u \ne v$, solve (6.7). Then, by the strict monotonicity, we obtain

$$0 < \langle A(u) - A(v), u - v \rangle = \langle f - f, u - v \rangle = 0,$$

which is a contradiction.

The warning about strictly monotone operators is same as with strictly convex functions. A non-coercive strictly convex function may not have a critical point. An equation with a non-coercive strictly monotone operator need not be solvable.

Remark 6.3

(i) The proof of Theorem 6.3 is constructive since the approximations represent the basis for many numerical methods.
(ii) Contrary to the approximating sequence from Theorem 6.2, which is convergent strongly due to the usage of the Banach Contraction Principle, the entire approximating sequence from the above theorem need not be even convergent weakly unless the uniqueness is guaranteed.
(iii) The assumption of reflexivity cannot be easily omitted due to the required weak compactness of the closed ball. On the other hand, the assumption of separability is not essential and the proof without it causes only some technical difficulties.

There are some exercises about the convergence of the approximating sequence in Theorem 6.3.

Exercise 6.2 Show that when A is strictly monotone, then the entire sequence $(u_n)_{n=1}^{\infty}$ constructed in the above proof is weakly convergent.

Exercise 6.3 Show that when A is strictly monotone and satisfies condition (S), then entire sequence $(u_n)_{n=1}^{\infty}$ constructed in the above proof is convergent strongly. Show that when operator A satisfies condition (S), then only certain subsequence $\left(u_{n_k}\right)_{k=1}^{\infty}$ converges in the norm.

We finish this section with investigating some properties of the inverse of operator A.

Theorem 6.4
Assume that $A : E \to E^*$ is radially continuous, strictly monotone, and coercive. Then A is invertible and $A^{-1} : E^* \to E$ is bounded, strictly monotone, and demicontinuous. When A satisfies additionally condition (S), then A^{-1} is continuous.

Proof By Corollary 6.1, we see that A is invertible. We divide the proof into some steps.

Operator A^{-1} is strictly monotone. Indeed, take $f, g \in E^*$, $f \neq g$. Then there are $u, v \in E, u \neq v$, such that $A(u) = f$, $A(v) = g$ for which we obtain

$$0 < \langle A(u) - A(v), u - v \rangle = \left\langle f - g, A^{-1}(f) - A^{-1}(g) \right\rangle.$$

Operator A^{-1} is bounded. Indeed, for any $M > 0$ take a bounded set

$$K \subset \left\{ f \in E^* : \|f\|_* \leq M \right\},$$

and consider

$$A^{-1}(K) = \{u \in E : A(u) = f \text{ for some } f \in K\}.$$

By the coercivity of A, there is a coercive function $\rho : [0, +\infty) \to \mathbb{R}$ such that for $u \in A^{-1}(K)$,

$$\|u\| \rho(\|u\|) \leq \langle A(u), u \rangle \leq \|A(u)\|_* \|u\| \leq M \|u\|.$$

Since ρ is coercive, this means that $\|u\| \leq M_1$ for some constant $M_1 > 0$.

Operator A^{-1} is demicontinuous. We use Lemma 3.6. Since A^{-1} is monotone, it suffices to prove that condition (ii) holds, i.e. for a fixed $u_0 \in E$, relation

$$\left\langle f_0 - g, u_0 - A^{-1}(g) \right\rangle \geq 0 \text{ for all } g \in E^* \tag{6.8}$$

implies that $A^{-1}(f_0) = u_0$. Suppose that (6.8) holds. Then putting $A^{-1}(g) = v$ we obtain from (6.8) the equivalent relation

$$\langle f - A(v), u_0 - v \rangle \geq 0 \text{ for all } v \in E. \tag{6.9}$$

Since A is radially continuous, by Lemma 3.6, condition (ii), we see that (6.9) implies that $A(u_0) = f$. Thus (6.8) implies that $A^{-1}(f) = u_0$. But this is equivalent, again by Lemma 3.6, to stating that A^{-1} is demicontinuous.

Operator A^{-1} is continuous. Finally, we assume that A satisfies condition (S) and assume that $f_n \to f_0$ in E^*. Put $u_n = A^{-1}(f_n)$ for $n \in \mathbb{N}$ and $u_0 = A^{-1}(f_0)$. Since A^{-1} is demicontinuous, we see that

$$u_n = A^{-1}(f_n) \rightharpoonup A^{-1}(f_0) = u_0.$$

Now it follows that

$$\langle A(u_n) - A(u_0), u_n - u_0 \rangle = \left\langle f_n - f_0, A^{-1}(f_n) - A^{-1}(f_0) \right\rangle \to 0 \text{ as } n \to +\infty.$$

Since A satisfies condition (S), we now see that

$$\left\| A^{-1}(f_n) - A^{-1}(u_0) \right\| = \|u_n - u_0\| \to 0 \text{ as } n \to +\infty.$$

But this means the continuity of A^{-1}.

Exercise 6.4 Examine the properties of the inverse of a uniformly monotone and a d-monotone operator.

Remark 6.4

The Minty Lemma 3.5 finds application in the alternative proof of the Browder–Minty Theorem, see Theorem 7.5 from [17].

6.3   Some Useful Corollaries

In Lemma 3.6, we introduced relation (iii), which is equivalent to the radial continuity under the assumptions that operator is monotone.

Definition 6.1 (Condition (M))

Let $A : E \to E^*$. If relations $u_n \rightharpoonup u_0$ in E, $A(u_n) \rightharpoonup f$ in E^* and

$$\limsup_{n \to +\infty} \langle A(u_n), u_n \rangle \leq \langle f, u_0 \rangle$$

imply that $A(u_0) = f$, then A is said to satisfy condition (M) or that A is of type (M).

Exercise 6.5 Let $A : E \to E^*$ be a demicontinuous operator of type $(S)_+$ and $C : E \to E^*$ be a compact operator. Show that $A + C$ is of type (M).

Exercise 6.6 Let $A : E \to E^*$ be an operator of type (M) and $C : E \to E^*$ a weakly continuous operator such that the function $x \mapsto \langle C(x), x \rangle$ is sequentially weakly lower semicontinuous. Check if the operator $A + C$ is of type (M).

Careful examination of the proof of Theorem 6.3 says that the monotonicity of the operator A is used in the following steps:

(i)   the continuity of A_n being a consequence of the fact that a radially continuous monotone operator is demicontinuous;
(ii)   local boundedness of A which is a consequence of the monotonicity; and
(iii)   the assertion (c) of Lemma 3.6 corresponding to the above mentioned condition (M).

The above made remarks allow us to proceed with a new version of the Browder–Minty Theorem.

Corollary 6.2

Assume that $A : E \to E^*$ is demicontinuous, bounded, and coercive and satisfies condition (M). Then for any $f \in E^*$, the set K of solutions to $A(u) = f$ is non-empty and bounded.

Proof The first part follows as in the proof of Theorem 6.3. Observe also that if the operator is bounded, then it is locally bounded. We invite the reader to repeat all necessary steps of the proof.

Remark 6.5

Observe that the above corollary is still valid when we replace the boundedness of operator A with its local boundedness.

Theorem 6.5 (Generalized Browder–Minty Theorem)
Assume that $B : E \to E^*$ is a coercive operator such that $B = A + T$, where $A : E \to E^*$ is monotone and radially continuous and $T : E \to E^*$ is strongly continuous. Then for any $f \in E^*$, the set K of solutions to $B(u) = f$ is non-empty and bounded.

Proof We will apply Corollary 6.2 in a form suggested by the above mentioned remark. We see that since A is monotone and radially continuous, it is also demicontinuous. Since T is strongly continuous, it follows that B is demicontinuous. Since T is strongly continuous, it follows that it is locally bounded. Take $u_n \rightharpoonup u_0$ in E such that $B(u_n) \rightharpoonup f$ in E^* and

$$\limsup_{n \to +\infty} \langle B(u_n), u_n \rangle \leq \langle f, u_0 \rangle. \tag{6.10}$$

We will show that $B(u_0) = f$, which means that condition (M) is satisfied and Corollary 6.2 applies. Since T is strongly continuous, it follows that $T(u_n) \to T(u_0)$, so that $A(u_n) \rightharpoonup f_1$, where $f_1 = f - T(u_0)$. Then (6.10) implies that

$$\limsup_{n \to +\infty} \langle A(u_n), u_n \rangle \leq \langle f_1, u_0 \rangle.$$

Now it follows from Lemma 3.6 that $A(u_0) = f_1$.

6.4   The Strongly Monotone Principle

Now we can prove the introductory existence result, namely Theorem 6.2 without the assumption of the Lipschitz continuity of operator A and also in a Banach space setting. This result, whose proof does not follow from the Banach Contraction Principle, is also known as the Strongly Monotone Principle.

Theorem 6.6 (Strongly Monotone Principle)

Assume that $A : E \to E^*$ is radially continuous and strongly monotone (with a constant $m > 0$). Then operator A is invertible and its inverse $A^{-1} : E^* \to E$ is Lipschitz continuous. If additionally operator A is Lipschitz continuous, then A^{-1} is strongly monotone.

Proof Since a strongly monotone operator is strictly monotone and since it satisfies condition (S), we see by Theorem 6.4 that operator A^{-1} is continuous. In what follows we take arbitrary $f, g \in E^*$ and $u, v \in E$ such that

$$A (u) = f, \ A (v) = g.$$

We have

$$\|f - g\|_* \|u - v\| \geq \langle A (u) - A (v), u - v \rangle \geq$$
$$m \|u - v\|^2 = m \left\| A^{-1} (f) - A^{-1} (g) \right\| \|u - v\|.$$

This means that A^{-1} is Lipschitz continuous.

Assume now that A is Lipschitz continuous with a constant L. Then

$$\frac{1}{L^2} \|f - g\|_*^2 = \frac{1}{L^2} \|A (u) - A (v)\|_*^2 \leq \|u - v\|^2.$$

It now follows from the above since A is strongly monotone that

$$\langle f - g, A^{-1} (f) - A^{-1} (g) \rangle = \langle A (u) - A (v), u - v \rangle \geq$$
$$m \|u - v\|^2 \geq \frac{m}{L^2} \|f - g\|_*^2.$$

Thus A^{-1} is strongly monotone.

6.5   Pseudomonotone Operators

Corollary 6.2 provides some conditions that are inspired by monotonicity but which do not involve the monotonicity directly. Therefore a new notion is now introduced:

Definition 6.2 (Pseudomonotone Operator)

We say that the operator $A : E \to E^*$ is pseudomonotone when the following implication holds:

if

$$u_n \rightharpoonup u_0 \text{ in } E$$

and if

$$\limsup_{n \to +\infty} \langle A\,(u_n)\,, u_n - u_0 \rangle \leq 0,$$

then

$$\liminf_{n \to +\infty} \langle A\,(u_n)\,, u_n - v \rangle \geq \langle A\,(u_0)\,, u_0 - v \rangle \text{ for all } v \in E.$$

The pseudomonotonicity differs from the monotonicity since it involves some type of continuity. We make this remark precise in the following lemma, which is about a type of a continuity of a pseudomonotone operator. The assumption that an operator is bounded is not very demanding.

Lemma 6.4

If an operator $A : E \to E^*$ is pseudomonotone and bounded, then it is demicontinuous.

Proof Take a sequence $(u_n)_{n=1}^{\infty} \subset E$ such that $u_n \to u_0$. Then sequences $(u_n)_{n=1}^{\infty}$ and $(A\,(u_n))_{n=1}^{\infty}$ are bounded. Thus there is a subsequence of $(u_n)_{n=1}^{\infty}$ not renumbered and an element $b \in E^*$ such that $A\,(u_n) \rightharpoonup b$. Since A is pseudomonotone, we see that for all $v \in E$,

$$\langle A\,(u_0)\,, u_0 - v \rangle \leq \liminf_{n \to +\infty} \langle A\,(u_n)\,, u_n - v \rangle = \langle b, u_0 - v \rangle\,.$$

Thus for all $h \in E$,

$$\langle A\,(u_0)\,, h - u_0 \rangle \leq \langle b, h - u_0 \rangle\,.$$

But this means that $A\,(u_0) = b$ and the assertion is proved.

The following exercises help us in understanding the above introduced notion together with connections to the already obtained existence results and to the monotonicity. At the same time some serve as sufficient conditions for pseudomonotonicity.

Exercise 6.7 Prove that the sum of two pseudomonotone operators forms a pseudomonotone operator.

Exercise 6.8 Show that if an operator $A : E \to E^*$ is monotone and hemicontinuous, then it is pseudomonotone.

Exercise 6.9 Show that if an operator $A : E \to E^*$ is strongly continuous, then it is pseudomonotone.

Exercise 6.10 Show that if an operator $A : E \to E^*$ is demicontinuous and satisfies condition $(S)_+$, then it is pseudomonotone.

Exercise 6.11 Show that if an operator is pseudomonotone, then it satisfies condition (M).

Exercise 6.12 Show that if E is a finite dimensional real Banach space and $A : E \to E^*$ is a continuous operator, then A is pseudomonotone.

Now we have the following version of the Browder–Minty Theorem involving pseudomonotone operators:

Theorem 6.7
Assume that $A : E \to E^*$ is pseudomonotone, bounded, and coercive. Then, for any fixed $g \in E^*$, the set K of solutions to $A(u) = g$ is non-empty.

Proof We see that assumptions of Corollary 6.2 are satisfied since a pseudomonotone and bounded operator is demicontinuous and it satisfies condition (M). This implies the assertion.

In a finite dimensional setting, we have the following consequence of the above result (which finds applications for difference equations, see for example [23]):

Theorem 6.8
Let E be a finite dimensional real Banach space, and let $A : E \to E^*$ be a continuous operator. Suppose that there exists a function $r : [0, +\infty) \to \mathbb{R}$ such that $\lim_{t \to +\infty} r(t) = +\infty$ and that inequality

$$(A(x), x) \geq r(\|x\|) \|x\|$$

holds for all $x \in E$. Then for each $g \in E^*$, equation $A(x) = g$ has at least one solution.

Exercise 6.13 Provide a detailed proof of the above theorem.

Exercise 6.14 Consider problem (1.15) from Sect. 1.3 with notation as in this section, which is why we do not repeat it. Assume additionally that for each $k \in \{1, 2, \ldots, N\}$, there exists a number $q_k > 0$ such that

$$f_k(x)x \geq q_k |x|^2 \text{ for } x \in \mathbb{R}.$$

Put

$$q = \min_{1 \le k \le N} \{q_k\}.$$

Using the above theorem, show that for any $\lambda \in (\frac{\lambda_N}{q}, +\infty)$, Eq. (1.15) has at least one solution.

6.6   The Leray–Lions Theorem

The next result is meant for applications to problems in which the so-called right hand side also depends on the derivative. Such a formulation makes the problem non-variational, i.e. there is no direct action functional whose critical points provide solutions to the given problem. This formulation of the Leray–Lions Theorem is coined after [14, 50].

Theorem 6.9 (Leray–Lions Theorem)

Let operator $T : E \to E^*$ be bounded, and coercive. Assume that there exists a mapping $\Phi : E \times E \to E^*$ such that

(i) $\Phi(u, u) = T(u)$ for every $u \in E$;

(ii) *for each $u \in E$, the mapping $\Phi(u, \cdot) : E \to E^*$ is bounded and hemicontinuous; for each $v \in E$ the mapping $\Phi(\cdot, v) : E \to E^*$ is bounded; moreover, for all $u, v \in E$, we have*

$$\langle \Phi(u, u) - \Phi(u, v), u - v \rangle \ge 0 \qquad (6.11)$$

(the so-called condition of monotonicity in the principal part);

(iii) *if $u_n \rightharpoonup u_0$ and if*

$$\lim_{n \to +\infty} \langle \Phi(u_n, u_n) - \Phi(u_n, u_0), u_n - u_0 \rangle = 0,$$

then we have

$$\Phi(u_n, w) \rightharpoonup \Phi(u_0, w) \text{ for arbitrary } w \in E; \text{ and}$$

(iv) *if $v \in E$, $u_n \rightharpoonup u_0$ in E and if $\Phi(u_n, v) \rightharpoonup z$ in E^*, then*

$$\lim_{n \to +\infty} \langle \Phi(u_n, v), u_n \rangle = \langle z, u_0 \rangle.$$

Then it follows that the equation

$$T(u) = f$$

has at least one solution $u_0 \in E$ for every fixed $f \in E^*$.

Proof For the proof of this result, we will use Theorem 6.7. Thus we need to show that operator T is pseudomonotone recalling that a bounded and pseudomonotone operator is necessarily demicontinuous. Thus we wish to demonstrate that if we select a sequence

$$u_n \rightharpoonup u_0 \text{ in } E$$

such that

$$\limsup_{n \to +\infty} \langle T(u_n), u_n - u_0 \rangle \leq 0, \tag{6.12}$$

then it follows that

$$\liminf_{n \to +\infty} \langle T(u_n), u_n - v \rangle \geq \langle T(u_0), u_0 - v \rangle \text{ for all } v \in E. \tag{6.13}$$

Since operator $\Phi(\cdot, u_0)$ is bounded, it follows that sequence $(\Phi(u_n, u_0))_{n=1}^{\infty}$ is bounded. Therefore, up to a subsequence, which we assume to be chosen and which we do not renumber, we have that $\Phi(u_n, u_0) \rightharpoonup z$, where $z \in E^*$. From (iv), we see that

$$\lim_{n \to +\infty} \langle \Phi(u_n, u_0), u_n \rangle = \langle z, u_0 \rangle. \tag{6.14}$$

Now we define numerical sequence $(x_n)_{n=1}^{\infty}$ as follows:

$$x_n = \langle \Phi(u_n, u_n) - \Phi(u_n, u_0), u_n - u_0 \rangle \text{ for } n \in \mathbb{N}.$$

Note that by (6.12), we obtain

$$\limsup_{n \to +\infty} \langle \Phi(u_n, u_n), u_n - u_0 \rangle \leq 0.$$

Moreover, from (6.14), we see that

$$\lim_{n \to +\infty} \langle \Phi(u_n, u_0), u_n - u_0 \rangle = 0. \tag{6.15}$$

This means that

$$\limsup_{n \to +\infty} x_n \leq 0.$$

By (6.11), we see that $x_n \geq 0$ for all $n \in \mathbb{N}$. Therefore $\lim_{n \to +\infty} x_n = 0$, which by (6.15) implies that

$$\lim_{n \to +\infty} \langle \Phi(u_n, u_n), u_n - u_0 \rangle = 0. \tag{6.16}$$

From (iii), it follows that

$$\Phi(u_n, w) \rightharpoonup \Phi(u_0, w) \text{ for arbitrary } w \in E, \tag{6.17}$$

and also from (iv), we have

$$\langle \Phi(u_n, w), u_n - u_0 \rangle \to 0 \text{ for arbitrary } w \in E. \tag{6.18}$$

Now we will make use of the monotonicity in the principal part again, i.e. inequality

$$\langle \Phi(u_n, u_n) - \Phi(u_n, w), u_n - w \rangle \geq 0 \text{ for all } w \in E.$$

We put $w = \theta v + (1 - \theta) u_0$ for $\theta \in (0, 1)$ and any $v \in E$. From the above, we have what follows

$$\theta \langle T(u_n), u_n - v \rangle = \theta \langle \Phi(u_n, u_n), u_n - v \rangle \geq$$
$$- (1 - \theta) \langle \Phi(u_n, u_n), u_n - u_0 \rangle + \langle \Phi(u_n, w), u_n - u_0 \rangle + \theta \langle \Phi(u_n, w), u_0 - v \rangle$$

for all $v \in E$. Observe that

$$\theta \liminf_{n \to +\infty} \langle \Phi(u_n, u_n), u_n - v \rangle \geq$$
$$- (1 - \theta) \liminf_{n \to +\infty} \langle \Phi(u_n, u_n), u_n - u_0 \rangle + \liminf_{n \to +\infty} \langle \Phi(u_n, w), u_n - u_0 \rangle +$$
$$\theta \liminf_{n \to +\infty} \langle \Phi(u_n, w), u_0 - v \rangle.$$

Then by (6.16), (6.17), and (6.18) after dividing by θ, we obtain

$$\liminf_{n \to +\infty} \langle T(u_n), u_n - v \rangle \geq \lim_{n \to +\infty} \langle \Phi(u_n, w), u_0 - v \rangle = \langle \Phi(u_0, w), u_0 - v \rangle$$

for all $v \in E$. Now letting $\theta \to 0$ by the hemicontinuity of the mapping $w \mapsto \Phi(u_0, w)$, we obtain

$$\liminf_{n \to +\infty} \langle T(u_n), u_n - v \rangle \geq \langle \Phi(u_0, u_0), u_0 - v \rangle$$

for all $v \in E$, which means that (6.13) is satisfied for a subsequence. Since it holds for each subsequence, it follows that (6.13) holds for the whole sequence, and therefore the theorem is proved.

Remark 6.6

We formulated the above theorem after [50] noting that in the source mentioned it is introduced a notion of an operator of the calculus of variation type. We did not include this notion formulating all required conditions directly in the theorem, and we relaxed the continuity assumption.

Remark 6.7

The conditions (i)–(iv) of the above theorem are somewhat not intuitive at the first glance. We will demonstrate how these conditions work by an application to boundary value problems for with the right hand side depending also on a derivative, see Sect. 9.7.

Normalized Duality Mapping

7

In the review article [49] one finds the following question: What can replace the inner product in non-Hilbert Banach spaces? The answer is around the notion of a duality mappings which we introduce following excerpts from [9, 19] and also some parts from [5] as far as the examples are concerned. We end this chapter with some information on a duality mapping relative to some function which as we shall see may better suit the structure of the space $W_0^{1,p}(0, 1)$. The theory concerning the notion just mentioned is given in detail in [9]. Here we provide only some necessary information following [13, 45]. The advanced reading may be found in [9, 24].

7.1 Introductory Notions and Properties

Recall that E is a real, separable, and reflexive Banach space. In this section we consider classical duality mapping together with its relation to the Riesz operator—in case we assume that we work in a Hilbert space. Monotonicity and continuity properties of the duality mapping are explored and some examples are given as well. We will need a special version of the Hahn-Banach Theorem (see Corollary 2.1.15 in [14]) and one of its consequences:

> **Theorem 7.1 (Hahn-Banach Theorem)**
> *Each linear and continuous functional f defined on a subspace F of E can be extended without changing its norm to the whole space; i.e. there is $f_1 \in E^*$ such that $\|f_1\| = \|f\|$ and $f_1(x) = f(x)$ for all $x \in F$.*

© The Author(s), under exclusive license to Springer Nature Switzerland AG 2021
M. Galewski, *Basic Monotonicity Methods with Some Applications*,
Compact Textbooks in Mathematics, https://doi.org/10.1007/978-3-030-75308-5_7

Proposition 7.1
For every $x \in E$ there exists $l \in E^*$ such that

$$\|l\|_* = 1 \text{ and } l(x) = \|x\|.$$

Now, we can proceed to the definition of a duality mapping for which we need an existence result:

Theorem 7.2
Assume that E^* is strictly convex. Then for each $x \in E$ there exists exactly one element $J(x) \in E^*$ such that

$$\langle J(x), x \rangle = \|x\|^2 = \|J(x)\|_*^2. \tag{7.1}$$

Proof Let $x \neq 0$. For $x = 0$ the result is obvious. On the space spanned by x we define a functional

$$f(tx) = t\|x\|^2, \text{ for } t \in \mathbb{R}.$$

The norm of this functional is easily calculated to be equal $\|x\|$. Then obviously (7.1) holds and by the Hahn-Banach Theorem we may extend f to space E preserving its norm. Suppose there are two distinct elements f_1, f_2 such that

$$\langle f_1, x \rangle = \langle f_2, x \rangle = \|x\|^2 = \|f_1\|_*^2 = \|f_2\|_*^2.$$

Then we have

$$\|f_1 + f_2\|_* \|x\| \geq \langle f_1 + f_2, x \rangle = \|f_1\|_* \|x\| + \|f_2\|_* \|x\|$$

and recalling that

$$\|f_1\|_* = \|f_2\|_*$$

we obtain

$$2\|f_1\|_* \leq \|f_1 + f_2\|_* < \frac{1}{2}\left(\|f_1\|_* + \|f_2\|_*\right).$$

Thus $f_1 = f_2$.

Definition 7.1 (Duality Mapping)
Assume that E^* is strictly convex. The operator $J : E \to E^*$ defined by (7.1) is called a duality mapping for space E.

Exercise 7.1 Check by a direct computation that the identity operator stands for a duality mapping for $L^2(0, 1)$.

Exercise 7.2 Assume that E^* is strictly convex. Show that the duality mapping J has the following property:

$$\langle J(x), x \rangle = \|x\| \, \|J(x)\|_* \text{ for each } x \in E.$$

Remark 7.1
If we drop the assumption that E^* is strictly convex, we may define the multivalued mapping $J : E \to 2^{E^*}$ as follows:

$$J(x) = \left\{ x^* \in E^* : \langle x^*, x \rangle = \|x^*\|_* \|x\| \right\}$$

which is called also a (multivalued) duality mapping. We direct the Reader to [9, 24] for a thorough treatment of this subject.

Now we proceed with investigation of the monotonicity and the continuity of a duality mapping.

Proposition 7.2
Assume that E^* is strictly convex. Then the duality mapping J is $d-$monotone, coercive, and demicontinuous.

Proof We first show that J is $d-$monotone. Let us take $u, v \in E$. Then we see

$$\langle J(u) - J(v), u - v \rangle =$$
$$\langle J(u), u \rangle - \langle J(u), v \rangle - \langle J(v), u \rangle + \langle J(v), v \rangle \geq$$
$$\|u\|^2 + \|v\|^2 - 2\|u\|\|v\| = (\|u\| - \|v\|)(\|u\| - \|v\|)$$

which means that J is $d-$monotone with respect to $\rho(x) = x$.

Now we show that J is demicontinuous. Let $u_n \to u_0$ in E. Then since $\|u_n\| = \|J(u_n)\|_*$, we see that the sequence $(J(u_n))_{n=1}^\infty$ is bounded and since E is reflexive, it contains a weakly convergent subsequence $\left(J\left(u_{n_j}\right) \right)_{j=1}^\infty$. Let its limit be denoted by $f_0 \in E^*$.

Then we have for any $y \in E$

$$\langle f_0, y \rangle = \lim_{j \to +\infty} \langle J \left(u_{n_j} \right), y \rangle \leq \lim_{j \to +\infty} \left\| J \left(u_{n_j} \right) \right\|_* \|y\| = \lim_{j \to +\infty} \left\| u_{n_j} \right\| \|y\|_* = \|u_0\| \|y\|_* .$$

Thus $\|f_0\|_* \leq \|u_0\|$. On the other hand we see that since $J \left(u_{n_j} \right) \rightharpoonup f_0$ and since $u_{n_j} \to u_0$ it follows

$$\langle f_0, u_0 \rangle = \lim_{j \to +\infty} \langle J \left(u_{n_j} \right), u_{n_j} \rangle = \lim_{j \to +\infty} \left\| u_{n_j} \right\|^2 = \|u_0\|^2 .$$

This means that $\|f_0\|_*^2 = \|u_0\|^2 = \langle f_0, u_0 \rangle$ and therefore $f_0 = J \left(u_0 \right)$. Since the above arguments hold for any weakly convergent subsequence of $(J \left(u_n \right))_{n=1}^{\infty}$, we see that $(J \left(u_n \right))_{n=1}^{\infty}$ converges weakly to $J \left(u_0 \right)$ and this implies the demicontinuity of J.

Finally we consider the coercivity of J. We recall that a $d-$monotone operator is coercive in case when $\rho \left(x \right) \to +\infty$ as $x \to +\infty$, see (3.1). But in our case we have $\rho \left(x \right) = x$. Hence the assertion follows.

The duality mapping is a potential operator as well.

Lemma 7.1

Assume that E^* is strictly convex. Then the duality mapping $J : E \to E^*$ is a potential operator with the potential $F : E \to \mathbb{R}$ given by

$$F \left(x \right) = \frac{1}{2} \|x\|^2 .$$

Proof Take any $u, v \in E$ and observe by the monotonicity of J that

$$\langle J \left(v \right), v - u \rangle \geq \langle J \left(u \right), v - u \rangle .$$

Thus putting for $h \in E, t \in \mathbb{R}$

$$v = u + th,$$

we see that

$$t \langle J \left(u + th \right), h \rangle \geq \frac{1}{2} \|u + th\|^2 - \frac{1}{2} \|u\|^2 \geq t \langle J \left(u \right), h \rangle .$$

Since J is demicontinuous and monotone, it is radially continuous. Thus

$$\langle J \left(u + th \right), h \rangle \to \langle J \left(u \right), h \rangle \text{ as } t \to 0.$$

Then we see that

$$\lim_{t \to 0} \frac{\frac{1}{2} \|u + th\|^2 - \frac{1}{2} \|u\|^2}{t} = \langle J(u), h \rangle .$$

Putting more restrictions on the space, we can have that the potential of the duality mapping J is not only Gâteaux differentiable, but also a C^1 mapping. This is equivalent to having J continuous. In what follows we are able to demonstrate when J is a homeomorphism.

Theorem 7.3

Assume that both E and E^* are uniformly convex. Then the duality mapping $J : E \to E^*$ is a homeomorphism. Moreover, $J^{-1} : E^* \to E$ is a duality mapping for space E^*.

Proof We need to prove that J is invertible, continuous and has a continuous inverse. Since a uniformly convex space is strictly convex, we see that J is strictly monotone. Using Proposition 7.2 and Theorem 6.4 we observe that J is invertible. By Lemma 3.3 it follows that a $d-$monotone operator in a uniformly convex space has property (S) and therefore J^{-1} is continuous. Using (7.1) we note that J^{-1} is a duality mapping for space E^*. Therefore repeating the arguments already used to operator J^{-1} we see that J is continuous as well.

Now, we turn to the case when E is a Hilbert space. We recall that by R we denoted the Riesz operator. Our aim now is to show that in fact $R^{-1} = J$. This is to no surprise if we again go carefully through the lines of the proof of Theorem 7.3.

Lemma 7.2

Let $J : E \to E^*$ be the duality mapping for a real, separable Hilbert space E. Then for all $x, y \in E$ we have

$$\langle J(x), y \rangle = (x, y)_E .$$

Proof Let $x \in E$ be fixed and consider functional $y \mapsto (x, y)_E$ which is obviously linear and continuous; moreover its norm is $\|x\|$. Thus to each x there corresponds an element $f \in E^*$ with the following property:

$$\langle f, y \rangle = (x, y)_E \quad \text{for all } y \in E.$$

Inserting $y = x$ in the above we have

$$\langle f, x \rangle = (x, x)_E = \|x\|^2 = \|f\|_*^2.$$

From the strict convexity of E and from Lemma 7.2 it follows that $f = J(x)$.

7.2   Examples of a Duality Mapping

We proceed with examples of duality mappings working for spaces considered in this text. These illustrate the monotonicity methods introduced. Several results collected here were mentioned earlier in a different context. Some serve as additional and direct exercises allowing the Reader for practising introduced tools. Thus is why we do not need to provide detailed proofs leaving the arguments to the Reader.

7.2.1   A Duality Mapping for $H_0^1(0, 1)$

Using Lemma 7.2 we see that

$$J : H_0^1(0, 1) \to H^{-1}(0, 1),$$

given by

$$\langle J(u), v \rangle = (u, v)_{H_0^1} = \int_0^1 \dot{u}(t)\, \dot{v}(t)\, dt \text{ for } u, v \in H_0^1(0, 1), \qquad (7.2)$$

is a duality mapping for $H_0^1(0, 1)$. Fixing $u \in H_0^1(0, 1)$ and calculating the norm of the functional

$$v \mapsto \langle J(u), v \rangle$$

we see that

$$\|J(u)\|_{H^{-1}} = \|u\|_{H_0^1}.$$

Summarizing, J is a duality mapping which is potential and that (7.2) stands for the derivative of its C^1 potential

$$u \mapsto \frac{1}{2} \|u\|_{H_0^1}^2.$$

The following exercise is of independent interest and provides a nice tool in investigating the formula for a duality mapping.

Exercise 7.3 Let H be a real separable Hilbert space with scalar product $(\cdot, \cdot)_H$ and let $J : H \to H^*$ be a duality mapping for H. Assume that Y is another real Hilbert space with scalar product $(\cdot, \cdot)_Y$ and that $L : H \to Y$ is a linear mapping such that

$$\|x\|_H = \|Lx\|_Y .$$

Show that then $J = L^*L$. Hint: use the uniqueness of a duality mapping and the following formula:

$$\left(L^*Lx, x\right)_H = (Lx, Lx)_Y = \|x\|_H^2 \text{ for } x \in H.$$

Exercise 7.4 Apply the above exercise to derive the formula for the duality mapping for $H_0^1 (0, 1)$. Hint: define

$$H = H_0^1 (0, 1) , \ Y = L^2 (0, 1) , \ L = \frac{d}{dt}.$$

Use also Example 3.5.

7.2.2 On a Duality Mapping for $L^p (0, 1)$

Let $p \geq 2$. We define $J : L^p (0, 1) \to L^q (0, 1)$ pointwisely for $u \neq 0$

$$J (u (\cdot)) = \|u\|_{L^p}^{2-p} |u (\cdot)|^{p-2} u (\cdot) \text{ a.e. on } [0, 1] \tag{7.3}$$

and $J (0) = 0$ or else by the following formula for $u \neq 0$

$$\langle J (u) , v \rangle = \|u\|_{L^p}^{2-p} \int_0^1 |u (t)|^{p-2} u (t) v (t) \, dt, \ v \in L^p (0, 1) .$$

Exercise 7.5 Show that J is well defined, i.e. for any $u \in L^p (0, 1)$ it holds

$$J (u (\cdot)) \in L^q (0, 1) .$$

Exercise 7.6 Show by a direct calculation that for any $u \in L^p (0, 1)$ it holds

$$\langle J (u) , u \rangle = \|u\|_{L^p}^2 = \|J (u)\|_{L^q}^2 .$$

By the above exercises and by Theorem 7.2 we see that operator J given by formula (7.3) stands for the duality mapping between $L^p (0, 1)$ and $L^q (0, 1)$.

7.2.3 On a Duality Mapping for $W_0^{1,p}(0,1)$

Let $p \geq 2$. It comes as no surprise, if we recall Lemma 3.2, that a duality mapping for $W_0^{1,p}(0,1)$ is an operator

$$J : W_0^{1,p}(0,1) \to W^{-1,q}(0,1)$$

given by

$$\langle J(u), v \rangle = \|u\|_{W_0^{1,p}}^{2-p} \int_0^1 |\dot{u}(t)|^{p-2}\, \dot{u}(t)\, \dot{v}(t)\, dt \tag{7.4}$$

for $u, v \in W_0^{1,p}(0,1)$, $u \neq 0$. We put $J(0) = 0$. We can calculate (which is again left as an exercise) that

$$\langle J(u), u \rangle = \|u\|_{W_0^{1,p}}^2 = \|J(u)\|_{W^{-1,q}}^2$$

for any $u \in W_0^{1,p}(0,1)$. Therefore by the uniqueness of a duality mapping we see that the assertion follows.

Remark 7.2

We see that for $p = 2$ formula (7.4) complies with the definition of the Laplacian.

7.3   On the Strongly Monotone Principle in Banach Spaces

In Sect. 6.1 we have considered the following equation:

$$A(u) = h, h \in E$$

with $A : E \to E$ strongly monotone and Lipschitz continuous and with E being a Hilbert space. The proof which followed by the Banach Contraction Principle provided also some numerical scheme. Later on we also considered the so-called Strongly Monotone Principle this time in a Banach space setting and proved via the Browder–Minty Theorem. Now we extend Theorem 6.2 to the case of mappings acting between E and E^* and use the Banach Contraction Principle as a main tool in the proof.

Theorem 7.4

Assume that E^* is strictly convex and $A : E \to E^*$ is Lipschitz continuous, i.e. there is $M > 0$ such that for all $u, v \in E$ we have

$$\|A(u) - A(v)\|_* \leq M \|u - v\|$$

and strongly monotone with a constant $m < M$, i.e. for $u, v \in E$ we have

$$\langle A(u) - A(v), u - v \rangle \geq m \|u - v\|^2 .$$

Then for each $h \in E^*$ equation

$$A(u) = h$$

has exactly one solution $u_0 \in E$.

Proof Let us fix $\varepsilon \in \left(0, \frac{2m}{M^2}\right)$. Let us define $B : E \to E$ by the following formula:

$$B(u) = u - \varepsilon J^{-1}(A(u) - h) \text{ for } u \in E.$$

Note that a fixed point of B is exactly the solution to equation $A(u) = h$. Since $\varepsilon \in \left(0, \frac{2m}{M^2}\right)$ the result easily follows by the Banach Contraction Principle.

Remark 7.3

Under assumptions of the above theorem for each $\varepsilon \in \left(0, \frac{2m}{M^2}\right)$ and each $v_0 \in E$ the element u_0 which solves equation $A(u) = h$ is a strong limit of the following iteration scheme:

$$J(v_n) = J(v_{n-1}) - \varepsilon(A(v_{n-1}) - h) \text{ for } n \in \mathbb{N}.$$

The following estimation holds for $n \in \mathbb{N}$

$$\|v_n - u_0\| \leq \frac{(k(t))^n t}{1 - k(t)} \|A(v_0) - h\|_* ,$$

where

$$k(t) = \sqrt{1 - 2mt + M^2 t^2} \text{ for } t \in \left(0, \frac{2m}{M^2}\right). \tag{7.5}$$

(continued)

Remark 7.3 (continued)
Function k defined by formula (7.5) attains infimum

$$k_0 = \sqrt{1 - (m/M)^2}$$

over $\left(0, \frac{2m}{M^2}\right)$ at $t_0 = \frac{m}{M^2}$. Observe also that $\sqrt{1 - 2mt + M^2 t^2} < 1$ for $t \in \left(0, \frac{2m}{M^2}\right)$.

Exercise 7.7 Provide the detailed proof of Theorem 7.4.

7.4   On a Duality Mapping Relative to a Normalization Function

In this section we will follow [13] in order to introduce the notion of duality mapping from E into E^* relative to a normalization function in a special case when E is a uniformly convex separable Banach space with a strictly convex dual. We assume such a setting in this section. We note that a more detailed treatment of this subject is to be found in [9]. The Reader will find a number of exercises to be considered with methods developed in Sect. 7.1.

Definition 7.2
A continuous function $\varphi : \mathbb{R}_+ \to \mathbb{R}_+$ is called a normalization function if it is strictly increasing, $\varphi(0) = 0$ and $\varphi(r) \to +\infty$ with $r \to +\infty$.

Definition 7.3
A duality mapping on E corresponding to a normalization function φ is an operator $J_\varphi : E \to 2^{E^*}$ such that for all $u \in E$ and $u^* \in J_\varphi(u)$

$$\left\|u^*\right\|_* = \varphi\left(\|u\|\right), \quad \langle u^*, u \rangle = \left\|u^*\right\|_* \|u\|. \tag{7.6}$$

From Proposition 7.1 it follows that there exists $l \in E^*$ such that

$$\|l\|_* = 1 \text{ and } l(u) = \|u\|.$$

Then $u^* = \varphi\left(\|u\|\right) l$ belongs to $J_\varphi(u)$. Hence $J_\varphi(u) \neq \emptyset$ for any $u \in E$.

Remark 7.4

We see that the duality mapping corresponding to the normalization function

$$\varphi(t) = t$$

is the already introduced normalized duality mapping.

Exercise 7.8 Prove that for any $x \in E$ the set $J_\varphi(x)$ is a singleton.

Exercise 7.9 Prove that functional $\psi : E \to \mathbb{R}$ defined by the following formula:

$$\psi(u) = \int_0^{\|u\|} \varphi(t)\, dt$$

is Gâteaux differentiable and convex. Calculate that J_φ stands for the Gâteaux derivative of functional ψ.

Exercise 7.10 Prove that $J_\varphi : E \to E^*$ is d−monotone and demicontinuous.

Exercise 7.11 Let $\varphi(t) = t^{p-1}$ for some $p \geq 2$. Show by a direct calculation using (7.6) that if $E = W_0^{1,p}(0, 1)$, then $J_\varphi = -\Delta_p$.

Exercise 7.12 Let $u \in W_0^{1,p}(0, 1)$. Using (7.6) and the above exercise calculate the norm of element $-\Delta_p u$ (the so-called dual norm of u).

On the Galerkin Method

8

In this chapter, we concentrate on the lines of proof of the Browder–Minty Theorem extracting from them the idea of Galerkin type approximation as given in [19] and also in [17]. Some information on this topic is also to be found in [18]. These concepts give rise to the numerical approximation of nonlinear problems that is vastly considered for example in [51] and also in [4]. What we present here is some theoretical introduction allowing the reader to pursue such topics in a different further study as well as the rich world of numerical procedures. We also mention connections to the Ritz type approximation, again providing only some sketch following this time [14] and [19].

8.1 Basic Notions and Results

The outline of this method has been used in the proof of Theorem 6.3. Now we fix some abstract scheme for which we also provide some new results. Recall that E is a separable, reflexive Banach space. Let us consider a complete system of linearly independent vectors $\{h_1, h_2, \ldots\} \subset E$. This means that

$$\overline{\text{span} \{h_1, h_2, \ldots\}} = E.$$

Let us define

$$E_n = \text{span} \{h_1, h_2, \ldots, h_n\}$$

for $n \in \mathbb{N}$. Note that E_n is considered with the inherited norm and being finite dimensional it is necessarily a real reflexive Banach space as well. Let

$$I_n : E_n \to E$$

denote the embedding of E_n into E. Let

$$I_n^* : E^* \to E_n^*$$

denote its adjoint. We see that for all $u \in E_n$, it follows

$$I_n u = u \text{ and } \|I_n u\| = \|u\|.$$

Then by Lemma 3.2, we see that for $n \in \mathbb{N}$, operator $A_n : E_n \to E_n^*$ defined by $A_n := I_n^* A I_n$ shares the monotonicity properties of $A : E \to E^*$, if any.

Observe also that for any $u_0 \in E$, there is a sequence $(u_n)_{n=1}^\infty$, $u_n \in E_n$, such that

$$\lim_{n \to +\infty} \|u_n - u_0\| = 0.$$

Definition 8.1 (Galerkin Solution)

Let $A : E \to E^*$ and let $g \in E^*$ be fixed. Element $u_n \in E$ given by

$$u_n = \sum_{j=1}^n a_j^n h_j, \text{ where } \left(a_1^n, \ldots, a_n^n\right) \in \mathbb{R}^n,$$

is called n-Galerkin type solution to

$$A(u) = g,$$

if vector $\left(a_1^n, \ldots, a_n^n\right) \in \mathbb{R}^n$ is a solution to

$$\langle A(u_n), h_j \rangle = \langle g, h_j \rangle \text{ for } j = 1, 2, \ldots, n. \tag{8.1}$$

Remark 8.1

Note that (8.1) can be written, using the remarks given prior to definition, as follows:

$$A_n(u_n) = g_n, \tag{8.2}$$

where $g_n = I_n^* g \in E_n^*$. Thus when we indicate that we seek for n-Galerkin type solution, we mean solution to (8.2).

Theorem 8.1 (The Galerkin Method)
Assume that operator $A : E \to E^*$ is radially continuous, strictly monotone, and coercive. Let $g \in E^*$ be fixed. Then for any $n \in \mathbb{N}$, there exists exactly one n-Galerkin type solution u_n to equation

$$A(u) = g \tag{8.3}$$

and moreover $u_n \rightharpoonup u_0$, where u_0 is the unique solution to (8.3).

Proof Since A is monotone and radially continuous, we know from Lemma 3.6 that it is demicontinuous. Now we see that A_n for $n \in \mathbb{N}$ is continuous, strictly monotone, and coercive as well. Then Corollary 6.1 provides that for any $n \in \mathbb{N}$, problem (8.2) has exactly one solution u_n. Using the same arguments, we see that u_0 solves (8.3). Since A is coercive, it follows for any $u \in E$ that

$$\|A(u)\|_* \geq \gamma(\|u\|)$$

for some coercive function $\gamma : [0, +\infty) \to [0, +\infty)$. Then for any $n \in \mathbb{N}$, we have

$$\|g\|_* \geq \|A(u_n)\|_* \geq \gamma(\|u_n\|).$$

This means that sequence $(u_n)_{n=1}^{\infty}$ is bounded. Let us take any weakly convergent subsequence $(u_{n_k})_{k=1}^{\infty}$ of $(u_n)_{n=1}^{\infty}$. We denote its limit by $\overline{u}$. Then, for any $v \in E_{n_k}$, we see by a direct calculation

$$\begin{aligned}
\langle g, v \rangle = \langle g, I_{n_k} v \rangle = \langle I_{n_k}^* g, v \rangle = \langle g_{n_k}, v \rangle = \\
\langle A_{n_k}(u_{n_k}), v \rangle = \langle I_{n_k}^* A(I_{n_k} u_{n_k}), v \rangle = \\
\langle A(I_{n_k} u_{n_k}), I_{n_k} v \rangle = \langle A(u_{n_k}), v \rangle.
\end{aligned}$$

This means, by the uniqueness of the weak limit, that $A(u_{n_k}) \rightharpoonup g$. Then we have from Lemma 3.6, condition (iii), that $A(\overline{u}) = g$. Since a solution to (8.3) is unique, we see that $\overline{u} = u_0$ and moreover that any weakly convergent subsequence of $(u_n)_{n=1}^{\infty}$ approaches u_0. Thus also $u_n \rightharpoonup u_0$.

The following characterization is now obtained easily:

Theorem 8.2
Assume $A : E \to E^*$ is radially continuous, strictly monotone, and coercive. Assume further that A satisfies condition (S). Let $g \in E^*$ be fixed. Then for any $n \in \mathbb{N}$, there

(continued)

Theorem 8.2 (continued)

exists exactly one n-Galerkin type solution u_n to (8.3) and $u_n \rightarrow u_0$, where u_0 is a unique solution to (8.3).

8.2   On the Galerkin and the Ritz Method for Potential Equations

Now we relate the method working for variational problems, namely the Ritz method, with the Galerkin method. This is why we assume that $A : E \rightarrow E^*$ is a potential operator with the potential $F : E \rightarrow \mathbb{R}$. Let $g \in E^*$ be fixed. For $n \in \mathbb{N}$, we use spaces E_n introduced above.

Definition 8.2 (Ritz Solution)

Let $n \in \mathbb{N}$. Element $u_n \in E_n$ is called the n-th Ritz solution for equation

$$A(u) = g, \tag{8.4}$$

if

$$J(u_n) = \min_{v \in E_n} J(v), \tag{8.5}$$

where functional $J : E \rightarrow \mathbb{R}$ is given by

$$J(u) = F(u) - \langle g, u \rangle.$$

In this section, by J, we mean the above introduced functional. Now we show that for potential problems, the Galerkin type and Ritz type approximations overlap. Monotonicity, as expected, is indispensable in the proof.

Theorem 8.3

Let $g \in E^*$ be fixed. Assume that $A : E \rightarrow E^*$ is a monotone and potential operator with the potential $F : E \rightarrow \mathbb{R}$. The element $u_n \in E_n$ is the n-th Ritz solution to problem (8.4) if and only if it is n-Galerkin type solution to (8.4).

Proof Assume that $u_n \in E_n$ is the n-th Ritz solution for equation (8.4). Then it holds by the **Fermat Rule** and by (8.5)

$$\langle A(u_n) - g, v \rangle = 0 \text{ for any } v \in E_n.$$

But this means that u_n is an n-Galerkin solution to (8.4). On the other hand if u_n is an n-Galerkin solution to (8.4), then by the convexity of F, we have

$$F(v) - F(u_n) + \langle g, v - u_n \rangle \geq \langle A(u_n) - g, v - u_n \rangle = 0$$

for any $v \in E_n$. Therefore $u_n \in E_n$ is the n-th Ritz solution for Eq. (8.4).

As an easy exercise, we leave to the reader the proof of the following result:

Theorem 8.4

Let $g \in E^*$ be fixed. Assume that $A : E \to E^*$ is a strictly monotone, coercive, and potential operator. Let moreover A satisfy condition (S). Then for each $n \in \mathbb{N}$, there exists exactly one n-th Ritz solution and $u_n \to u_0$ in E, where u_0 is a unique solution to (8.4).

Some Selected Applications

9

Collecting various examples and notes from the sources which we mentioned, we give some applications of the main theoretical results, also for the second order Dirichlet boundary value problem. The results in this chapter are derived following various sources among which we mention: [14,17,18,32,37,58]. Using the theory of Sobolev spaces given in [1] one may easily shift the examples from Sects. 9.5–9.7 to the field of partial differential equations. The regularity may not be reached as well as some proofs would need adjustment in order to replace the uniform convergence with the convergence in a suitable L^p space. One may see how this works, for example, in [12, 13] as well as in [50, 54]. Such applications are also mentioned in [14]. The book devoted to partial equations with various monotonicity approaches is, for example, [43]. An introduction to application of variational inequalities via the monotonicity theory is in [36]. Advanced applications for potential equations are to be found in [7], while a modern approach toward nonlinear problems is to be found in [41, 47]. Applications for Dirichlet problems on fractal domains are given in [20].

9.1 On Nonlinear Lax-Milgram Theorem and the Nonlinear Orthogonality

We apply Theorem 6.2 to a version of nonlinear Lax-Milgram Theorem which may be viewed as a type of the nonlinear orthogonality. The classical Lax-Milgram Theorem is also derived. In this section we assume that E is a real, separable Hilbert space. We recall that:

Definition 9.1 (Bounded Bilinear Form)
Bounded bilinear form on E is a functional

$$a : E \times E \to \mathbb{R}$$

© The Author(s), under exclusive license to Springer Nature Switzerland AG 2021
M. Galewski, *Basic Monotonicity Methods with Some Applications*,
Compact Textbooks in Mathematics, https://doi.org/10.1007/978-3-030-75308-5_9

fulfilling the following conditions:

(i) for all $u, v, w \in E, \alpha, \beta \in \mathbb{R}$ it holds

$$a\left(\alpha u + \beta v, w\right) = \alpha a\left(u, w\right) + \beta a\left(v, w\right),$$

$$a\left(w, \alpha u + \beta v\right) = \alpha a\left(w, u\right) + \beta a\left(w, v\right);$$

(ii) there exists a constant $d > 0$ such that for all $u, v \in E$:

$$\left|a\left(u, v\right)\right| \leq d\left\|u\right\|\left\|v\right\|.$$

Definition 9.2

(i) Form $a : E \times E \to \mathbb{R}$ is called symmetric if and only if

$$a\left(u, v\right) = a\left(v, u\right)$$

for all $u, v \in E$.
(ii) Form a is called positive if and only if

$$a\left(u, u\right) \geq 0$$

for all $u \in E$; a is called strictly positive if and only if

$$a\left(u, u\right) > 0$$

for all $u \in E, u \neq 0$.
(iii) Form a is called strongly positive if and only if there exists a constant $c > 0$ such that

$$a\left(u, u\right) \geq c\left\|u\right\|^2$$

for all $u \in E$.

Exercise 9.1   Prove that a bounded bilinear form $a : E \times E \to \mathbb{R}$ is a continuous functional.

Example 9.1   The Reader is invited to prove that the scalar product in any real Hilbert space is a bounded symmetric strongly positive bilinear form. In particular in $H_0^1\left(0, 1\right)$ we have

$$a\left(u, v\right) = \int_0^1 \dot{u}\left(t\right) \dot{v}\left(t\right) dt \text{ for } u, v \in H_0^1\left(0, 1\right)$$

and we see that $c = 1$ and $d = 1$.

Here is the nonlinear orthogonality principle.

Theorem 9.1   (Lax-Milgram Theorem)
Assume that:

(i) $b : E \to \mathbb{R}$ is a linear continuous functional ($b \in E^*$);
(ii) $a : E \times E \to \mathbb{R}$ is a mapping such that for each $w \in E$

$$v \mapsto a(w, v)$$

is a linear continuous functional;
(iii) there are positive constants $m > 0$ and $L > 0$ such that for $u, v, w \in E$

$$m \|u - v\|^2 \le a(u, u - v) - a(v, u - v)$$

and

$$|a(u, w) - a(v, w)| \le L \|u - v\| \|w\|.$$

Then there is exactly one $u \in E$ which solves equation

$$a(u, v) = \langle b, v \rangle. \tag{9.1}$$

Proof By assumption (i) and by the Riesz Representation Theorem for each $w \in E$ there is an element $A(w)$ such that

$$a(w, v) := (A(w), v)_E$$

for all $v \in E$. We have thus defined an operator A from E into E. From assumption (iii) it follows that

$$(A(u) - A(v), u - v)_E \ge m \|u - v\|^2$$

and

$$\|A(u) - A(v)\| = \sup_{\|z\| \le 1} |(A(u) - A(v), z)_E| \le \sup_{\|z\| \le 1} L \|u - v\| \|z\| \le L \|u - v\|.$$

This means that A is strongly monotone and Lipschitz continuous. Again by the Riesz Representation Theorem there is an element $h \in E$ such that

$$\langle b, v \rangle = (h, v)_E \quad \text{for all } v \in E.$$

Thus (9.1) is equivalent to

$$A(u) = h$$

and since all assumptions of Theorem 6.2 are satisfied, we reach the assertion.

A direct relation of the above result to bilinear forms justifies that we call it a nonlinear orthogonality principle. Now we proceed to some version of the Lax-Milgram Theorem (see also Corollary 5.8 from [5]) which finds many applications in the solvability of boundary value problems for differential equations (partial including):

Theorem 9.2
Assume that $a : E \times E \to \mathbb{R}$ is a bounded and strongly positive bilinear form. Then for any given element $\varphi \in E^*$ there exists a unique u_0 such that

$$a(u_0, h) = \langle \varphi, h \rangle \text{ for all } h \in E. \tag{9.2}$$

Moreover, if additionally form a is symmetric, then u_0 is the unique minimizer to the following functional:

$$J(u) = \frac{1}{2}a(u, u) - \langle \varphi, u \rangle, \ J : E \to \mathbb{R}.$$

Remark 9.1
The direct application of Theorem 2.22 to the above is as follows under the additional assumption that a is symmetric. It is easy to calculate that J is C^1. Functional J is coercive since for all $u \in E$ it holds:

$$J(u) = \frac{1}{2}a(u, u) - \langle \varphi, u \rangle \geq \frac{1}{2}c \|u\|^2 - \|\varphi\|_* \|u\|.$$

Calculating the second variation we see that

$$J^{(2)}(u; h, h) = a(h, h) \geq \frac{1}{2}c \|h\|^2 > 0$$

for all $h \neq 0, h \in E$ and all $u \in E$. Summarizing, functional J is strictly convex, sequentially weakly lower semicontinuous, and coercive. Therefore it has exactly argument of a minimum u_0 which satisfies (9.2).

Exercise 9.2 Prove that functional J given in the Lax-Milgram Theorem is C^1 and check the related formulas for both the derivative and the second Gâteaux variation.

9.2   On a Certain Converse of the Lax-Milgram Theorem

Now we turn to investigating the relations between bilinear forms and continuous linear mappings and monotonicity notions. This section has been prepared with some ideas from [48]. Let E be a real, separable Hilbert space and let $A : E \to E$ be a linear continuous operator. Define $a : E \times E \to \mathbb{R}$ by

$$a(u, v) = (Au, v)_E \text{ for } u, v \in E.$$

Then it is left as an exercise to prove that a is a bilinear form. Operator A is called positive, strictly positive, strongly positive, or symmetric if a has the corresponding property.

Exercise 9.3 Let $A : E \to E$ be a linear continuous operator. Prove that:

  (i) a is positive if and only if A is monotone;
 (ii) a is strictly positive if and only if A is strictly monotone;
(iii) a is strongly positive if and only if A is strongly monotone.

The following exercise is necessary for the next result.

Exercise 9.4 Assume that $A : E \to E$ is a positive and invertible linear operator. Define $h : E \to \mathbb{R}$ by

$$h(x) = \frac{1}{2}(Ax, x)_E.$$

Show that $h^* : E \to \mathbb{R}$ reads as follows:

$$h^*(v) = \frac{1}{2}\left(A^{-1}v, v\right)_E.$$

Lemma 9.1

Assume that $A : E \to E$ is a positive and invertible linear operator. Then A is coercive (in particular, A is positive definite). As a matter of fact,

$$(Ax, x)_E \geq \left\|A^{-1}\right\|^{-1} \|x\|^2 \text{ for all } x \in E.$$

Proof Define

$$h\left(x\right) = \frac{1}{2}\left(Ax, x\right)_E.$$

From Exercise 9.4 it follows

$$h^*\left(v\right) = \frac{1}{2}\left(A^{-1}v, v\right)_E.$$

Then from the Fenchel-Young Inequality we see what follows for $v = \alpha x$ and $\alpha = \left\|A^{-1}\right\|^{-1}$

$$\left(Ax, x\right)_E + \left(A^{-1}v, v\right)_E \geq 2\alpha\left(x, x\right)_E.$$

Hence we have using the estimation

$$\left(A^{-1}v, v\right)_E \leq \left\|A^{-1}v\right\| \|v\| \leq \alpha \|x\|^2,$$

(following from the Schwarz Inequality and the definition of α) that

$$\left(Ax, x\right)_E \geq \left\|A^{-1}\right\|^{-1} \|x\|^2.$$

Exercise 9.5 Show that $\left\|A^{-1}\right\|^{-1}$ determined above is the optimal (i.e. largest possible) coercivity coefficient. Hint: assume that there is another such constant and argue by contradiction.

Remark 9.2

If a is a symmetric, continuous, and strongly positive bilinear form, then we can conclude from the Lax-Milgram Theorem that it defines a continuous linear operator A which is invertible and positive. Lemma 9.1 and the exercise which follows imply a quantitative converse: if A is strictly positive (or positive) and invertible, then a is coercive, and, moreover, the best possible coercivity coefficient of a is $\left\|A^{-1}\right\|^{-1}$.

9.3   Applications to the Differentiability of the Fenchel-Young Conjugate

We are now interested in the differentiability of the Fenchel-Young conjugate of a differentiable and convex functional. Such result is related to the content of Theorem 5.2 but we now use existence theorems also in the proofs which we did not previously.

Theorem 9.3

Assume that operator $A : E \to E^*$ is strictly monotone, coercive, and potential. Then it is invertible and $A^{-1} : E^* \to E$ is strictly monotone, bounded and potential. Relation

$$F(v) = \int_0^1 \langle A(sv), v \rangle \, ds$$

defines the potential for A and for any $x^* \in E^*$ it holds

$$F^*(x^*) = F^*(0) + \int_0^1 \langle x^*, A^{-1}(sx^*) \rangle \, ds, \; F^*(0) = -F\left(A^{-1}(0)\right). \quad (9.3)$$

Proof By Lemma 5.4 a monotone and potential operator is demicontinuous. Thus we see by Theorem 6.4 that A is invertible and A^{-1} is bounded, strictly monotone and demicontinuous. By Theorem 5.2 we know that F^* is Gâteaux differentiable with A^{-1} being its Gâteaux derivative which means that first relation in (9.3) holds. Using relation (iii) from Theorem 5.2 we get

$$F\left(A^{-1}(0)\right) + F^*(0) = 0.$$

Hence the second relation (9.3) is also satisfied.

The above results are given in the language of potential operators. We can reformulate it from the perspective of C^1 functionals and prove using critical point theory. Nevertheless, for the fact that F^* is C^1 it seems we will need some results from monotonicity theory pertaining to the usage of condition (S).

Corollary 9.1

Assume that $F : E \to \mathbb{R}$ is a C^1 functional which is strictly convex and strongly coercive, i.e.

$$\lim_{\|u\| \to +\infty} \frac{F(u)}{\|u\|} = +\infty. \quad (9.4)$$

Then its Fenchel-Young conjugate $F^* : E^* \to \mathbb{R}$ has the following form:

$$F^*(x^*) = \left\langle x^*, \left(F'\right)^{-1}(x^*) \right\rangle - F\left(\left(F'\right)^{-1}(x^*)\right). \quad (9.5)$$

Functional F^* is differentiable and $\left(F'\right)^{-1}$ is a derivative of F^*. If mapping $F' : E \to E^*$ satisfies additionally condition (S), then F^* is a C^1 functional as well.

Proof From Theorem 9.3 we know that F' is invertible. We will prove now that formula (9.5) holds. Let x^* be a fixed element of E^*. Indeed, by definition it follows

$$F^* \left(x^* \right) = - \inf_{x \in E} \left\{ F \left(x \right) - \left\langle x^*, x \right\rangle \right\}.$$

Note that functional $g : E \rightarrow \mathbb{R}$ defined by

$$g \left(x \right) = F \left(x \right) - \left\langle x^*, x \right\rangle$$

is continuous and strictly convex as a sum of a convex and a strictly convex functionals. Hence it is sequentially weakly lower semicontinuous. Since g is by (9.4) also coercive, it has a unique minimizer x_0 which satisfies the **Fermat Rule**

$$F' \left(x_0 \right) - x^* = 0$$

understood in the sense of space E^*. Due to the invertibility of F' we have

$$x_0 = \left(F' \right)^{-1} \left(x^* \right).$$

Thus we see that formula (9.5) holds.

In order to show that $\left(F' \right)^{-1}$ is continuous we use Theorem 6.4. We must show that F' is coercive. Indeed, the convexity of F yields that for all $x \in E$:

$$F \left(x \right) - F \left(0 \right) \leq \left\langle F' \left(x \right), x \right\rangle.$$

Then for $x \neq 0$

$$\frac{F \left(x \right) - F \left(0 \right)}{\|x\|} \leq \frac{\left\langle F' \left(x \right), x \right\rangle}{\|x\|}.$$

Therefore F' is coercive. Since F is convex it follows that F' is monotone. Since it is assumed that F' satisfies condition (S), we finally obtain the assertion that its inverse $\left(F' \right)^{-1}$ is continuous by Theorem 6.4.

9.4   Applications to Minimization Problems

We consider again the existence of a solution to the following problem:

$$A \left(u \right) = h,$$

where $A : E \rightarrow E^*$ is a potential mapping and $h \in E^*$ is fixed. We can treat the equation considered as a Euler–Lagrange equation to a suitable action functional. Such a functional exists by the assumption that A is potential. We have already obtained some preliminary result in Lemma 5.3. Now, we wish to prove the Browder–Minty Theorem via Theorem 2.22 in the potential case under some additional assumption satisfied in most cases. Next, we include the strongly continuous perturbations of A as well in order to obtain a counterpart of Theorem 6.5.

Lemma 9.2

Assume that operator $A : E \to E^*$ is monotone, bounded, coercive, and potential. Let $h \in E^*$ be fixed. Then there is at least one solution to equation

$$A(u) = h \qquad (9.6)$$

which is unique in case A is strictly monotone.

Proof By Lemma 5.4 we see that functional $J : E \to \mathbb{R}$ defined by

$$J(u) = \int_0^1 \langle A(tu), u \rangle \, dt - \langle h, u \rangle \qquad (9.7)$$

is sequentially weakly lower semicontinuous and differentiable in the sense of Gâteaux. The Gâteaux derivative of J at any fixed $u \in E$ reads

$$\left\langle J'(u), v \right\rangle = \langle A(u) - h, v \rangle \quad \text{for all } v \in E.$$

In order to apply Theorem 2.22 we observe that by Lemma 5.5 functional J is coercive.

Our next aim is to pursue the use of monotonicity results in putting some new light on variational problems. We start with the following lemma which says that any sequence which approaches strongly the solution of the Euler–Lagrange equation of type (9.6) with a monotone and potential left hand side is in fact a minimizing sequence. The existence of such a solution is, however, assumed.

Lemma 9.3

Assume that $A : E \to E^*$ is monotone and potential. Let moreover $h \in E^*$ be fixed and let u_0 be any solution to (9.6). Let $J : E \to \mathbb{R}$ be given by (9.7). If $(u_n)_{n=1}^\infty \subset E$ is such that $u_n \to u_0$, then

$$J(u_n) \to J(u_0) = \min_{u \in E} J(u).$$

Proof From Lemma 5.3 it follows that

$$J (u_0) = \min_{u \in E} J (u) .$$

Moreover, by Lemma 6.3 it follows that operator A is locally bounded. Since J is convex and Gâteaux differentiable we see by Theorem 3.1 that for any $n \in \mathbb{N}$ it holds

$$J (u_0) \leq J (u_n) \leq J (u_0) - \langle A (u_n) , u_0 - u_n \rangle \leq$$
$$J (u_0) + \| A (u_n) \|_* \| u_n - u_0 \| .$$

From the above estimation it follows that $J (u_n) \to J (u_0)$ as $n \to +\infty$.

As we easily see, the monotonicity of A is crucial in the above proof. Thus it is interesting if there is a counterpart of the above result without monotonicity. Moreover, it is interesting if we can say anything about improving the convergence of a minimizing sequence in Theorem 2.22, namely if these can converge strongly. For the proof of the next result we need the Ekeland Variational Principle which cite after [16], Theorem 4.4:

Theorem 9.4 (Ekeland Variational Principle-Differentiable Form)
Let $J : E \to \mathbb{R}$ be a Gâteaux differentiable functional which is bounded from below. Then there exists a minimizing sequence $(u_n)_{n=1}^{\infty}$ consisting of almost critical points, i.e. such a sequence that

$$J (u_n) \to \inf_{u \in E} J (u) \text{ and } J^{'} (u_n) \to 0 \text{ (in } E^*).$$

We have the following two recent results from [22] which utilize usage of monotonicity methods in exploration of the variational ones.

Theorem 9.5
Let operator $A : E \to E^*$ be monotone, bounded, potential, and satisfying condition (S). Let operator $T : E \to E^*$ be potential and strongly continuous. Let also $A + T$ be coercive. Then there is a solution u_0 to

$$A (u) + T (u) = h \tag{9.8}$$

which minimizes functional $J : E \to \mathbb{R}$ defined by

$$J (u) = \int_0^1 \langle A (su) , u \rangle \, ds + \int_0^1 \langle T (su) , u \rangle \, ds - \langle h, u \rangle . \tag{9.9}$$

(continued)

Theorem 9.5 (continued)

Moreover, there is a (non-constant) minimizing sequence $(u_n)_{n=1}^{\infty} \subset E$, $u_n \to u_0$ *(strongly) in E and such that*

$$J(u_n) \to \inf_{u \in E} J(u) \text{ and } J'(u_n) \to 0 \text{ (in } E^*). \tag{9.10}$$

Proof We see that functional $v \mapsto \int_0^1 \langle T(sv), v \rangle \, ds$ is sequentially weakly continuous. Indeed, since T is strongly continuous, it follows for any weakly convergent sequence $v_n \rightharpoonup v_0$ that

$$\int_0^1 \langle T(sv_n), v_n \rangle \, ds \to \int_0^1 \langle T(sv_0), v_0 \rangle \, ds. \tag{9.11}$$

Hence functional $v \mapsto \int_0^1 \langle T(sv), v \rangle \, ds$ is sequentially weakly lower semicontinuous. Now, by Lemma 5.5 functional J is coercive, as a potential of a coercive mapping. It is also sequentially weakly lower semicontinuous and Gâteaux differentiable. Therefore by Theorem 2.22 functional J has the argument of a minimum u_0 which is a critical point, i.e. a solution to (9.8).

Obviously J is bounded from below. Theorem 9.4 shows that there is a minimizing sequence $(u_n)_{n=1}^{\infty}$ such that conditions (9.10) are satisfied. By the coercivity of J we can assume, taking a subsequence if necessary, that $u_n \rightharpoonup u_0$. Moreover, by Theorem 9.4, we see that the minimizing sequence can be chosen so that it consists of almost critical points.

We prove that $u_n \to u_0$. Since T is strongly continuous, we see that $T(u_n) \to T(u_0)$. Since

$$J'(u_n) = A(u_n) + T(u_n) - h,$$

it follows from condition $J'(u_n) \to 0$ and from $T(u_n) \to T(u_0)$, that $A(u_n) \to f$ in E^*, where $f \in E^*$ is some fixed element. We will show that $f = A(u_0)$. From the monotonicity of A we have

$$\langle A(u_n) - A(v), u_n - v \rangle \geq 0 \text{ for any } v \in E.$$

Passing to a limit we see that

$$\langle f - A(v), u_0 - v \rangle \geq 0 \text{ for any } v \in E.$$

The above relation by Lemma 3.6 means that $f = A(u_0)$. Thus $A(u_n) \to A(u_0)$. Therefore we see that

$$\langle A(u_n) - A(u_0), u_n - u_0 \rangle \to 0.$$

Since condition (S) is satisfied, we obtain that $\|u_n - u_0\| \to 0$ as well.

In the setting of the above theorem, we also have a counterpart of Lemma 9.3.

Proposition 9.1

Assume that operator $A : E \to E^*$ is monotone and potential. Assume that operator $T : E \to E^*$ is potential and strongly continuous. Let u_0 be a solution to (9.8) minimizing functional (9.9), i.e.

$$A(u_0) + T(u_0) = 0 \text{ and } J(u_0) = \min_{u \in E} J(u).$$

If $(u_n)_{n=1}^{\infty} \subset E$ is such that $u_n \to u_0$, then

$$J(u_n) \to J(u_0) = \min_{u \in E} J(u).$$

Proof By convexity of the potential of A we see that

$$\int_0^1 \langle A(su_n), u_n \rangle \, ds \leq \int_0^1 \langle A(su_0), v \rangle \, ds - \langle A(u_n), u_0 - u_n \rangle.$$

Note that since A is monotone and potential it is demicontinuous, see Lemma 5.4. Since A is demicontinuous and monotone, it is locally bounded, see Proposition 3.3. This means that there is a constant $M > 0$ such that for $n \in \mathbb{N}$:

$$\|A(u_n)\|_* \leq M$$

since $(u_n)_{n=1}^{\infty}$ is convergent. Moreover, since T is strongly continuous we see that (9.11) holds. Therefore we have for $n \in \mathbb{N}$:

$$J(u_0) \leq \int_0^1 \langle T(su_n), u_n \rangle \, ds + \int_0^1 \langle A(su_n), u_n \rangle \, ds \leq$$
$$\int_0^1 \langle T(su_n), u_n \rangle \, ds + \int_0^1 \langle A(su_0), v \rangle \, ds - \langle A(u_n), u_0 - u_n \rangle \leq$$
$$\int_0^1 \langle T(su_n), u_n \rangle \, ds + \int_0^1 \langle A(su_0), v \rangle \, ds + M \|u_n - u_0\|.$$

From the above estimation it follows that $J(u_n) \to J(u_0)$ as $n \to +\infty$.

9.5   Applications to the Semilinear Dirichlet Problem

Starting from this section we turn to the application to boundary value problems. We begin with a linear problem corresponding to (1.1) with $a = 0$ leaving a more general case to Sect. 9.7. Let us fix $g \in L^2(0, 1)$ and let us consider the solvability in the space $H_0^1(0, 1)$ of the following problem:

$$\begin{cases} -\ddot{u}(t) = g(t), \text{ for } a.e.\ t \in (0, 1) \\ u(0) = u(1) = 0. \end{cases} \tag{9.12}$$

The solvability of (9.12) in a space $H_0^1 (0, 1)$ means that we seek the so-called weak solution. The weak solution to (9.12) is a such function $u \in H_0^1 (0, 1)$ that

$$\int_0^1 \dot{u} (t) \, \dot{v} (t) \, dt = \int_0^1 g (t) \, v (t) \, dt \text{ for all } v \in H_0^1 (0, 1) . \qquad (9.13)$$

We observed already by the du Bois-Reymond Lemma that under the above assumptions on g any weak solution (9.12) belongs to $H_0^1 (0, 1) \cap H^2 (0, 1)$. This means that the equation in understood a.e. on $[0, 1]$.

We wish to proceed with standard application of the monotonicity approach toward solvability of nonlinear problems which can be summarized as follows:

- determine the form of the possibly nonlinear operator from formula (9.13) defining the weak solution;
- check that the operator has any kind of monotonicity notions which we have introduced or else that it is a perturbation of some monotone operator;
- check the continuity of the operator which should be at least radial and the strong continuity of the perturbation;
- check the coercivity;
- investigate the uniqueness.

We arrive at the following result:

Theorem 9.6

For any $g \in L^2 (0, 1)$ *there exists exactly one* $u \in H_0^1 (0, 1) \cap H^2 (0, 1)$ *which solves* (9.12).

Proof We apply Theorem 7.3 and the results from Sect. 7.2.1. Alternatively, we apply the Strongly Monotone Principle (Theorem 6.6) following the calculations form Example 3.2.

Remark 9.3

We see that problem (9.12) is not only uniquely solvable but that the solution depends continuously on a parameter g. More precisely speaking, consider a convergent sequence of parameters $(g_n)_{n=1}^\infty \subset L^2 (0, 1)$ with $\lim_{n \to +\infty} g_n = g_0$. Then to each element g_n there corresponds a unique solution u_n to (9.12) with $g = g_n$. Moreover, since $-\Delta$ is a homeomorphism, we see that $\lim_{n \to +\infty} u_n = u_0$, where u_0 is a solution corresponding to g_0.

In order to formulate a more general Dirichlet problem (in fact a system of nonlinear equations when $N > 1$) with nonlinear term depending also on a function u, we assume that

A1 $f : [0, 1] \times \mathbb{R}^N \to \mathbb{R}^N$ *is an* L^2-*Carathéodory function with* $f(t, 0) = 0$ *for a.e.* $t \in [0, 1]$; $g \in L^2(0, 1)$, $g \neq 0$.

Now we consider the existence and uniqueness results for the following problem which is the counterpart of (1.1)

$$\begin{cases} -\ddot{u}(t) + f(t, u(t)) = g(t), \text{ for } a.e.\ t \in (0, 1) \\ u(0) = u(1) = 0. \end{cases} \tag{9.14}$$

We say that a function $u \in H_0^1(0, 1)$ is a weak solution of (9.14) if the following equality:

$$\int_0^1 \dot{u}(t)\,\dot{v}(t)\,dt + \int_0^1 f(t, u(t))v(t)dt = \int_0^1 g(t)\,v(t)\,dt \tag{9.15}$$

holds for every $v \in H_0^1(0, 1)$. We see that with assumption **A1** any solution is non-zero. We prove this assertion by a direct calculation assuming to the contrary. Using the du Bois-Reymond Lemma, Lemma 2.5, we see that any weak solution is in fact a classical one, i.e. $u \in H_0^1(0, 1) \cap H^2(0, 1)$. The assumption concerning the nonlinear term which we need in order to apply Theorem 6.6 is:

A2   *for a.e.* $t \in [0, 1]$ *the operator* $x \mapsto f(t, x)$ *is monotone on* $\mathbb{R}^N$.

Let us define operator $A : H_0^1(0, 1) \to H^{-1}(0, 1)$ as follows:

$$\langle A(u), v \rangle = \int_0^1 \dot{u}(t)\,\dot{v}(t)\,dt + \int_0^1 f(t, u(t))v(t)\,dt \tag{9.16}$$

for $u, v \in H_0^1(0, 1)$. Observe that operator (9.16) is well defined. We have already noted that $A_1 = -\Delta$ is well defined. As for the operator

$$\langle A_2(u), v \rangle = \int_0^1 f(t, u(t))v(t)\,dt \tag{9.17}$$

we fix $u \in H_0^1(0, 1)$ and note that, by the Schwarz Inequality and by the Poincaré Inequality, we obtain:

$$\int_0^1 f(t, u(t))v(t)\,dt \leq \sqrt{\int_0^1 f^2(t, u(t))dt}\sqrt{\int_0^1 v^2(t)\,dt} \leq c\,\|v\|_{H_0^1}$$

for all $v \in H_0^1(0, 1)$. Here $c := \sqrt{\int_0^1 f^2(t, u(t))dt/\pi}$.

Let us also define a continuous linear functional $g^* : H_0^1(0, 1) \to \mathbb{R}$, i.e. $g^* \in H^{-1}(0, 1)$, as follows:

$$\langle g^*, v \rangle = \int_0^1 g(t) v(t) \, dt.$$

Exercise 9.6 Verify that $g^* \in H^{-1}(0, 1)$.

Therefore problem (9.14) is equivalent to

$$A(u) = g^*.$$

Since this is in fact (9.15), we observe that abstract equation $A(u) = g^*$ provides weak solutions to (9.14). We have the following result:

Theorem 9.7

Assume that conditions A1, A2 are satisfied. Then problem (9.14) has exactly one nontrivial solution $u \in H_0^1(0, 1) \cap H^2(0, 1)$.

Proof We see that $A : H_0^1(0, 1) \to H^{-1}(0, 1)$ is a (possibly non-linear) strongly monotone operator. Indeed, for $u, v \in H_0^1(0, 1)$ we see

$$\langle A(u) - A(v), u - v \rangle =$$
$$\int_0^1 (\dot{u}(t) - \dot{v}(t))^2 \, dt + \int_0^1 (f(t, u(t)) - f(t, v(t))) (u(t) - v(t)) \, dt \geq$$
$$\| u - v \|_{H_0^1}^2 .$$

Operator A_1 is continuous. The continuity of A_2 follows by the generalized Krasnosel'skii Theorem 2.12. Indeed, given a sequence $(u_n)_{n=1}^\infty \subset H_0^1(0, 1)$ such that $u_n \to u_0$ we see that $u_n \rightrightarrows u_0$ which means that $(u_n)_{n=1}^\infty$ is uniformly bounded by some $d > 0$. Since f is L^2–Carathéodory function, we see that there is some $f_d \in L^2(0, 1)$ such that

$$|f(t, u_n(t))| \leq f_d(t) \text{ for } a.e. \ t \in [0, 1] \text{ and all } n \in \mathbb{N}.$$

By the application of generalized Krasnosel'skii Theorem 2.12 we see that

$$f(\cdot, u_n(\cdot)) \to f(\cdot, u_0(\cdot)) \text{ in } L^2(0, 1).$$

Summarizing, the existence and uniqueness result follows by the application of Theorem 6.6. Reasoning by a contradiction we see that a solution is necessarily nontrivial.

Exercise 9.7 Show that when $f : [0, 1] \times \mathbb{R}^N \to \mathbb{R}^N$ is an L^1–Carathéodory function then operator A_2 given by (9.17) is well defined and continuous.

We can replace **A2** with the following relaxed monotone condition:

A3   *there exists a constant* $0 < a_2 < \pi^2$ *such that for a.e.* $t \in [0, 1]$ *it holds for any* $x, y \in \mathbb{R}^N$

$$(f(t, x) - f(t, y), x - y) \geq -a_2 |x - y|.$$

Exercise 9.8 Prove Theorem 9.7 under assumptions **A1, A3**.

Apart from the Strongly Monotone Principle, Theorem 6.6, which requires the nonlinear term f to be at least relaxed monotone, we may apply Theorem 6.5 in which such restrictions do not appear. We impose instead some growth condition on f:

A4   *there exists a constant* $a_1 < \pi^2$ *and a function* $b_1 \in L^1(0, 1)$ *such that*

$$(f(t, x), x) \geq -a_1 |x|^2 + b_1(t) \tag{9.18}$$

for all $x \in \mathbb{R}^N$ *and a.e.* $t \in [0, 1]$.

Note that condition (9.18) is required in order to make sure that operator A defined by (9.16) is coercive. This is why there appears restriction on constant a_1 which in fact follows from the application of the Poincaré Inequality. We have the following result about the existence:

Theorem 9.8
*Assume that conditions **A1, A4** are satisfied. Then problem (9.14) has at least one nontrivial solution* $u \in H_0^1(0, 1) \cap H^2(0, 1)$.

Proof We see that operator A_2, see (9.17), is strongly continuous. The arguments follow as in the proof of Theorem 9.7 since given a sequence $(u_n)_{n=1}^{\infty} \subset H_0^1(0, 1)$ such that $u_n \rightharpoonup u_0$ we see that $u_n \to u_0$ in $C[0, 1]$ as well.

Therefore, in order to apply Theorem 6.5 we yet need to show that A given by (9.16) is coercive. Indeed, we have for $u \in H_0^1(0, 1)$ by the Poincaré and Sobolev Inequalities

$$\langle A(u), u \rangle \geq \int_0^1 |\dot{u}(t)|^2 \, dt - a_1 \frac{1}{\pi^2} \int_0^1 |\dot{u}(t)|^2 \, dt - \|b_1\|_{L^1} \|u\|_{H_0^1} \geq$$
$$\left(1 - \frac{a_1}{\pi^2}\right) \|u\|_{H_0^1}^2 - \|b_1\|_{L^1} \|u\|_{H_0^1}.$$

Then

$$\frac{\langle A(u), u \rangle}{\|u\|} \geq \left(1 - \frac{a_1}{\pi^2}\right) \|u\|_{H_0^1} - \|b_1\|_{L^1}$$

and it follows that A is coercive. The application of Lemma 2.5 shows again that any solution u belongs to $H_0^1 (0, 1) \cap H^2 (0, 1)$.

Exercise 9.9   Consider the following Dirichlet problem:

$$\begin{cases} -\ddot{u} (t) + f (u (t)) = g (t), & \text{for } a.e.\ t \in (0, 1) \\ u (0) = u (1) = 0, \end{cases}$$

where $g \in L^2 (0, 1)$ and where $f : \mathbb{R}^N \to \mathbb{R}^N$ is a continuous function such that

$$(f (x), x) \geq 0 \text{ for all } x \in \mathbb{R}^N.$$

Prove that the above problem has at least one classical solution. Hint: use Theorem 6.5.

9.5.1   Examples and Special Cases

Now we are concerned with a simpler problem

$$\begin{cases} -\ddot{u} (t) + f (u (t)) = g (t), & \text{for } a.e.\ t \in (0, 1) \\ u (0) = u (1) = 0, \end{cases} \tag{9.19}$$

where $g \in L^2 (0, 1)$, $f : \mathbb{R} \to \mathbb{R}$ is of one of the following types:

  (i) $f (x) = cx$, where $c > 0$ is fixed;
 (ii) $f (x) = cx^3$, where $c > 0$ is fixed;
(iii) f is a continuous nondecreasing function;
(iv) f is an arbitrary continuous function such that

$$\liminf_{|x| \to +\infty} f (x) \operatorname{sgn} (x) > -\infty. \tag{9.20}$$

In cases (i)–(iii) operator A_2 defined by (9.17) is monotone which in examples (i)–(ii) is proved by a direct calculation. We included example (iii) already considered above so that to compare it with (iv). As for (iv) we get rid of monotonicity at the expense of additional assumption (9.20) which is imposed so that to have coercivity of an operator A defined exactly as before.

Exercise 9.10   Consider the solvability of (9.19) under assumptions (i), (ii), (iv).

Apart from the above applications for the standard Laplacian, we can make use of Example 3.2 and replace (9.14) with the following:

$$\begin{cases} -\frac{d}{dt} \left(b (t) \frac{d}{dt} u (t) \right) + f (t, u (t)) = g (t), & \text{for } a.e.\ t \in (0, 1) \\ u (0) = u (1) = 0, \end{cases} \tag{9.21}$$

where $b \in L^\infty (0, 1)$, $b(t) \geq m > 0$ for a.e. $t \in [0, 1]$, $g \in L^1 (0, 1)$. In this case
the application of the du Bois-Reymond Lemma leads to conclusion that function
$t \mapsto b(t) \dot{u}(t)$ has a derivative a.e. on $[0, 1]$ which means that the solution does
not necessarily belong to $H^2 (0, 1)$. Operator $A : H_0^1 (0, 1) \to H^{-1} (0, 1)$ is now
defined as follows:

$$\langle A(u), v \rangle = \int_0^1 b(t) \dot{u}(t) \dot{v}(t) \, dt + \int_0^1 f(t, u(t)) v(t) \, dt.$$

Exercise 9.11 Find growth conditions on nonlinear term f under which problem (9.21) has:

(i) exactly one weak solution;
(ii) at least one weak solution.

9.6   Applications to Problems with the Generalized p–Laplacian

Let $p > 2$ and let us first consider the p–Laplacian

$$-\Delta_p : W_0^{1,p} (0, 1) \to W^{-1,q} (0, 1)$$

defined by

$$\langle -\Delta_p (u), v \rangle = \int_0^1 |\dot{u}(t)|^{p-2} \dot{u}(t) \dot{v}(t) \, dt$$

for $u, v \in W_0^{1,p} (0, 1)$. Now we can summarize what can be said about this operator
with the tools which we have developed so far.

Theorem 9.9
Operator

$$-\Delta_p : W_0^{1,p} (0, 1) \to W^{-1,q} (0, 1)$$

is well defined, potential, continuous, bounded, and uniformly monotone. Moreover,
$-\Delta_p$ *is a homeomorphism.*

Proof We recall that $-\Delta_p$ is bounded, potential, and continuous, see Proposition 5.1. We
observe that from the well known inequality

$$\left(|\eta|^{p-2} \eta - |\xi|^{p-2} \xi, \eta - \xi \right) \geq (1/2)^p |\eta - \xi|^p$$

valid for all $\eta, \xi \in \mathbb{R}^N$, we get for $u, v \in W_0^{1,p}(0, 1)$:

$$
\begin{aligned}
&\left\langle -\Delta_p(u) - \left(-\Delta_p(v)\right), u - v \right\rangle = \\
&\int_0^1 \left(|\dot{u}(t)|^{p-2} \dot{u}(t) - |\dot{v}(t)|^{p-2} \dot{v}(t), \dot{u}(t) - \dot{v}(t) \right) dt \geq \\
&(1/2)^p \int_0^1 |\dot{u}(t) - \dot{v}(t)|^p \, dt = \rho \left(\|u - v\|_{W_0^{1,p}} \right) \|u - v\|_{W_0^{1,p}},
\end{aligned}
$$

where

$$
\rho(x) = (1/2)^p \, x^{p-1} \text{ for } x \geq 0.
$$

Thus operator $-\Delta_p$ is uniformly monotone. Therefore it is also coercive and satisfies the property (S). By the strict monotonicity of $-\Delta_p$, we see that this operator is invertible. Thus by Theorem 6.4 operator $\left(-\Delta_p\right)^{-1}$ is continuous which means that $-\Delta_p$ defines a homeomorphism between $W_0^{1,p}(0, 1)$ and $W^{-1,q}(0, 1)$.

Now we proceed to consider the Dirichlet problem related to the $p-$Laplacian. We assume that

A5 $\varphi : [0, 1] \times \mathbb{R}_+ \to \mathbb{R}_+$ *is a Carathéodory function and there exists constant* $M > 0$ *such that*

$$
\varphi(t, x) \leq M
$$

for a.e. $t \in [0, 1]$ *and for all* $x \in \mathbb{R}_+$; *there exists a constant* $\gamma > 0$ *such that*

$$
\varphi(t, x) x - \varphi(t, y) y \geq \gamma (x - y)
$$

for all $x \geq y \geq 0$ *and for a.e.* $t \in [0, 1]$.

We consider a more general nonlinear operator

$$
A_1 : W_0^{1,p}(0, 1) \to W^{-1,q}(0, 1),
$$

given by

$$
\langle A_1(u), v \rangle = \int_0^1 \varphi\left(t, |\dot{u}(t)|^{p-1}\right) |\dot{u}(t)|^{p-2} \dot{u}(t) \dot{v}(t) \, dt \text{ for } u, v \in W_0^{1,p}(0, 1).
$$

$$
(9.22)
$$

Concerning equations involving the above introduced operator we will follow the scheme developed for (9.12), i.e. we will start from the problem with fixed right hand side and next we will proceed with nonlinear problems:

Proposition 9.2
*Assume that **A5** holds. Then operator A_1 defined by (9.22) is continuous and potential
with the potential $F : W_0^{1,p}(0, 1) \to \mathbb{R}$ defined by*

$$F(u) = \int_0^1 \int_0^{|\dot{u}(t)|} \varphi\left(t, s^{p-1}\right) s^{p-1} ds\, dt.$$

*Moreover, A_1 is coercive and $d-$monotone with respect to $\rho(x) = \gamma x^{p-1}$. Addition-
ally, operator A_1 is invertible and its inverse A_1^{-1} is continuous.*

Proof Using (3.8) and relation $1/p + 1/q = 1$ we obtain that

$$\int_0^1 \left|\varphi\left(t, |\dot{u}(t)|^{p-1}\right)\right|^q \left(|\dot{u}(t)|^{p-2}|\dot{u}(t)|\right)^q dt \le M^q \int_0^1 |\dot{u}(t)|^p.$$

But this means that A_1 is well defined. Applying Theorem 2.12 we see that A_1 is continuous.

From Theorem 3.3 we know that A_1 is $d-$monotone with respect to $\rho(x) = \gamma x^{p-1}$ and
by Theorem 5.1 it is potential. Exercise 3.25 provides that it is coercive. Since $W_0^{1,p}(0, 1)$
is uniformly convex (and therefore strictly convex) we are able to conclude by Remark 3.2
that A_1 is strictly monotone. By Lemma 3.3 it follows that operator A satisfies property (S).
Therefore by Theorem 6.4 operator A_1 is invertible and its inverse A_1^{-1} is continuous.

As announced above we begin with the solvability of a problem with a fixed right
hand side.

Proposition 9.3
*Assume that $g \in L^q(0, 1)$ is fixed and that condition **A5** holds. Then problem*

$$\begin{cases} -\frac{d}{dt}\left(\varphi\left(t, \left|\frac{d}{dt}u(t)\right|^{p-1}\right)\left|\frac{d}{dt}u(t)\right|^{p-2}\frac{d}{dt}u(t)\right) = g(t), \text{ for a.e. } t \in (0, 1), \\ u(0) = u(1) = 0 \end{cases}$$

has a unique weak solution $u \in W_0^{1,p}(0, 1)$, i.e. for any $v \in W_0^{1,p}(0, 1)$ it holds

$$\int_0^1 \varphi\left(t, |\dot{u}(t)|^{p-1}\right)|\dot{u}(t)|^{p-2}\dot{u}(t)\dot{v}(t)\, dt = \int_0^1 g(t)v(t)\, dt. \tag{9.23}$$

Proof We write Eq. (9.23) as follows:

$$A_1(u) = g^*, \tag{9.24}$$

where the linear and bounded functional $g^* : W_0^{1,p}(0, 1) \to \mathbb{R}$ is given by

$$g^*(v) = \int_0^1 g(t) v(t) \, dt, \tag{9.25}$$

and where A_1 is defined by (9.22). By Proposition 9.2 we see that Theorem 6.4 can be applied to problem (9.24). This finishes the proof.

Exercise 9.12 Prove that g^* defined above belongs to $W^{-1,q}(0, 1)$, i.e. that g^* is a linear and continuous functional over $W_0^{1,p}(0, 1)$.

In order to consider a problem with a nonlinear right hand side, we need to make some assumptions:

A6 $f : [0, 1] \times \mathbb{R}^N \to \mathbb{R}^N$ is an L^1–Carathéodory function with $f(t, 0) = 0$ for a.e. $t \in [0, 1]$ and assume that $g \in L^q(0, 1)$, $g \neq 0$.
A7 for a.e. $t \in [0, 1]$ operator $x \mapsto f(t, x)$ is monotone on $\mathbb{R}^N$.

Now we can consider the existence and also uniqueness result for the following problem:

$$\begin{cases} -\frac{d}{dt}\left(\varphi\left(t, \left|\frac{d}{dt}u(t)\right|^{p-1}\right)\left|\frac{d}{dt}u(t)\right|^{p-2}\frac{d}{dt}u(t)\right) + f(t, u(t)) = g(t), \ \text{for a.e. } t \in (0, 1) \\ u(0) = u(1) = 0. \end{cases}$$
$$\tag{9.26}$$

We say that a function $u \in W_0^{1,p}(0, 1)$ is a weak solution of (9.26) if for all $v \in W_0^{1,p}(0, 1)$ it holds

$$\int_0^1 \varphi\left(t, |\dot{u}(t)|^{p-1}\right)|\dot{u}(t)|^{p-2}\dot{u}(t)\dot{v}(t)\,dt + \int_0^1 f(t, u(t))v(t)\,dt = \int_0^1 g(t)v(t)\,dt.$$

We see that with assumption **A6** any solution is non-zero which we prove by a direct calculation assuming to the contrary.

Let us define operators $A, A_2 : W_0^{1,p}(0, 1) \to W^{-1,q}(0, 1)$ as follows:

$$\langle A_2(u), v \rangle = \int_0^1 f(t, u(t))v(t)\,dt, \tag{9.27}$$

$$\langle A(u), v \rangle = \langle A_1(u), v \rangle + \langle A_2(u), v \rangle$$

for $u, v \in W_0^{1,p}(0, 1)$.

Exercise 9.13 Show that A_2 is well defined, i.e. it sends points from $W_0^{1,p}(0, 1)$ into functionals working on $W_0^{1,p}(0, 1)$.

With g^* given by (9.25), we see that problem (9.26) is equivalent to the following abstract equation:

$$A(u) = g^*.$$

This in turn complies with the definition of the weak solution. We have the following result:

Theorem 9.10

Assume that conditions A5–A7 are satisfied. Then problem (9.26) has exactly one nontrivial solution.

Proof Recall that A_1 is strictly monotone and continuous. Note that for any $u, v \in W_0^{1,p}(0, 1)$

$$\int_0^1 (f(t, u(t)) - f(t, v(t)))(u(t) - v(t))\, dt \ge 0$$

which implies that A_2 is monotone. From Proposition 9.2 we see that A is strictly monotone and coercive. In order to prove that operator A_2 is continuous we use Theorem 2.12. Indeed, since a convergent sequence $(u_n)_{n=1}^\infty \subset W_0^{1,p}(0, 1)$ is uniformly bounded by some $d > 0$, then by **A6** there is some function $f_d \in L^1(0, 1)$ such that

$$|f(t, x)| \le f_d(t) \text{ for a.e. } t \in [0, 1] \text{ and } x \in [-d, d].$$

But this leads the strong continuity of A_2.

The application of Theorem 6.4 finishes the proof of the existence and the uniqueness of the solution. Reasoning by a contradiction we see that this solution is necessarily nontrivial.

Apart from Theorem 6.4 we may apply Theorem 6.5 for which require some growth condition on f instead of assumption **A7**:

A8   *there exists a constant $a_1 < \gamma$ such that*

$$(f(t, x), x) \ge -a_1 |x|^{p-1}$$

for all $x \in \mathbb{R}^N$ and for a.e. $t \in [0, 1]$.

We have the following result:

Theorem 9.11

Assume that conditions A5, A6, A8 are satisfied. Then problem (9.26) has at least one nontrivial solution.

Proof As in the proof of Theorem 9.10 we see that operator A_2 defined by (9.27) is strongly continuous. Hence in order to apply Theorem 6.5 we need to prove that A is coercive. Indeed, for any $u \in W_0^{1,p}(0,1)$ we obtain

$$\langle A(u), u \rangle \geq \gamma \int_0^1 |\dot{u}(t)|^p \, dt - a_1 \int_0^1 |u(t)|^p \, dt \geq (\gamma - a_1) \|u\|_{W_0^{1,p}}^p.$$

Therefore the coercivity of A follows and by Theorem 6.5 we obtain the assertion.

Exercise 9.14 Consider the following Dirichlet problem:

$$\begin{cases} -\frac{d}{dt}\left(\varphi\left(t, |\frac{d}{dt}u(t)|^{p-1}\right)|\frac{d}{dt}u(t)|^{p-2}\frac{d}{dt}u(t)\right) + f(u(t)) = g(t), & \text{for } a.e.\ t \in (0,1) \\ u(0) = u(1) = 0, \end{cases}$$

(9.28)

where $g \in L^q(0,1)$ and where $f : \mathbb{R}^N \to \mathbb{R}^N$ is a continuous function such that

$$(f(x), x) \geq 0 \text{ for all } x \in \mathbb{R}^N.$$

Using Theorem 6.5 prove that (9.28) has at least one weak solution.

Exercise 9.15 Assume that $f : \mathbb{R} \to \mathbb{R}$ is continuous and nondecreasing. Prove that (9.28) has exactly one weak solution.

Exercise 9.16 Check whether assumption **A8** can be replaced with the following: *there exists a constant $a_1 < \gamma$ and a function $b_1 \in L^q(0,1)$ such that*

$$(f(t,x), x) \geq -a_1 |x|^{p-1} + b(t) |x|$$

for all $x \in \mathbb{R}^N$ and for a.e. $t \in [0,1]$.

9.7 Applications of the Leray–Lions Theorem

The Leray–Lions Theorem is about the existence of solutions to second order nonlinear problems involving also first order derivatives. We may at last study problem corresponding to (1.1) with a nonlinear term as well. Let us fix $g \in L^2(0,1)$ and $a \in L^\infty(0,1)$ such that there is a constant $a_1 < \pi$

$$a(t) \geq a_1 > 0 \text{ for a.e. } t \in [0,1].$$

(9.29)

Assume that $f : [0, 1] \times \mathbb{R} \to \mathbb{R}$ is a Carathéodory function. We apply Theorem 6.9 in order to consider the existence of a solution for the following problem:

$$\begin{cases} -\ddot{u}(t) + f(t, u(t)) + a(t)\dot{u}(t) = g(t), & \text{for } a.e. \ t \in (0, 1), \\ u(0) = u(1) = 0 \end{cases} \tag{9.30}$$

under the assumptions:

A9 *there are constants $c > 0$, $m > 1$ and a function $f_0 \in L^1(0, 1)$ that such that*

$$|f(t, x)| \leq c\left(f_0(t) + |x|^m\right)$$

for a.e. $t \in [0, 1]$ and all $x \in \mathbb{R}$;
A10 *for a.e. $t \in [0, 1]$ and all $x \in \mathbb{R}$ it holds*

$$xf(t, x) \geq 0.$$

We look for weak solutions of (9.30), i.e. such functions $u \in H_0^1(0, 1)$ that

$$\int_0^1 \dot{u}(t)\dot{v}(t)\,dt + \int_0^1 f(t, u(t))v(t)\,dt + \int_0^1 a(t)\dot{u}(t)v(t)\,dt = \int_0^1 g(t)v(t)\,dt$$

for all $v \in H_0^1(0, 1)$. This formula suggests as usual that we should consider operator

$$T : H_0^1(0, 1) \to H^{-1}(0, 1)$$

given by the following formula for $u, v \in H_0^1(0, 1)$:

$$\langle T(u), v \rangle = \int_0^1 \dot{u}(t)\dot{v}(t)\,dt + \int_0^1 f(t, u(t))v(t)\,dt + \int_0^1 a(t)\dot{u}(t)v(t)\,dt.$$

Exercise 9.17 When a is constant show that operator

$$\langle T_1(u), v \rangle = \int_0^1 \dot{u}(t)\dot{v}(t)\,dt + \int_0^1 a\dot{u}(t)v(t)\,dt \ \text{for } u, v \in H_0^1(0, 1)$$

is strongly monotone and next apply Theorem 6.5 in order to examine the existence of a weak solution to (9.30). Hint: consult Example 3.7 in showing that T_1 is strongly monotone.

Since T_1 is not strongly monotone when a is some function satisfying (9.29) we will apply Theorem 6.9 in order to reach the existence result. Hence we must properly define mapping Φ and demonstrate that all assumptions (i)–(iv) of Theorem 6.9 are satisfied.

We proceed now with the following definitions which are introduced in order to separate the effects of higher and lower derivatives:

$$g^* : H_0^1(0, 1) \to \mathbb{R}, \, g^*(u) = \int_0^1 g(t) \, w(t) \, dt,$$

$$B : H_0^1(0, 1) \to H^{-1}(0, 1), \, \langle B(v), w \rangle = \int_0^1 \dot{v}(t) \, \dot{w}(t) \, dt,$$

$$G : H_0^1(0, 1) \to H^{-1}(0, 1), \, \langle G(u), w \rangle = \int_0^1 (f(t, u(t)) + a(t) \dot{u}(t)) \, w(t) \, dt,$$

for $u, v, w \in H_0^1(0, 1)$. Now we put

$$\Phi : H_0^1(0, 1) \times H_0^1(0, 1) \to H^{-1}(0, 1)$$

by the following formula:

$$\langle \Phi(u, v), w \rangle = \langle B(v), w \rangle + \langle G(u), w \rangle \tag{9.31}$$

for $u, v, w \in H_0^1(0, 1)$.

We start with lemma summarizing some obvious properties of operator T. The Reader is invited to provide detailed calculations as an exercise.

Lemma 9.4

Assume that conditions (9.29) and A9, A10 are satisfied. Then operator T is bounded, continuous and coercive.

Proof Observe that

$$\langle T(u), v \rangle = \langle B(u) + G(u), v \rangle$$

for all $u, v \in H_0^1(0, 1)$. We know that $B = -\Delta$ is bounded and continuous. As for the continuity of G we argue using the generalized Krasnosel'skii Theorem, see Theorem 2.12. By the same arguments it follows that G is bounded. We easily observe that T is coercive. Indeed, note by **A10** that $\langle G(u), u \rangle \geq 0$ for all $u \in H_0^1(0, 1)$. Moreover the following estimation holds due to (9.29) and the Poincaré Inequality

$$\int_0^1 a(t) \dot{u}(t) u(t) \, dt \geq -\frac{a_1}{\pi} \|u\|_{H_0^1}^2 \quad \text{for all } u \in H_0^1(0, 1).$$

By a direct calculation we see that operator B is coercive. Therefore we have the assertion of the lemma satisfied.

Lemma 9.5

*Assume that conditions **A9, A10** are satisfied and that operator Φ is defined by (9.31). Then conditions (i)–(iv) from Theorem 6.9 are satisfied.*

Proof By a direct calculation we see that $\Phi(u, u) = T(u)$ for every $u \in H_0^1(0, 1)$, so (i) holds. The first part of condition (ii) follows from Lemma 9.4.

For all $u, v \in H_0^1(0, 1)$ we directly calculate that

$$\langle \Phi(u, u), u - v \rangle = \langle B(u), u - v \rangle + \langle G(u), u - v \rangle,$$

$$\langle \Phi(u, v), u - v \rangle = \langle B(v), u - v \rangle + \langle G(u), u - v \rangle.$$

The above relations and the monotonicity of B imply that

$$\langle \Phi(u, u) - \Phi(u, v), u - v \rangle \geq 0$$

for all $u, v \in H_0^1(0, 1)$. This is why the condition of monotonicity in the principal part holds and we have the assumption (ii) satisfied.

Applying similar calculations and the fact operator B has the property (S), we see that if $u_n \rightharpoonup u_0$ and if

$$\lim_{n \to +\infty} \langle \Phi(u_n, u_n) - \Phi(u_n, u_0), u_n - u_0 \rangle = \langle B(u_n) - B(u_0), u_n - u_0 \rangle = 0,$$

it follows that $u_n \to u_0$. This provides that condition (iii) holds.

We finally prove that condition (iv) holds. Take $v \in H_0^1(0, 1)$, $u_n \rightharpoonup u_0$ and assume that $\Phi(u_n, v) \rightharpoonup z$ which means that

$$\langle B(v), u_n \rangle + \langle G(u_n), w \rangle \to 0 \text{ for all } w \in H_0^1(0, 1).$$

Since $u_n \rightharpoonup u_0$ in $H_0^1(0, 1)$ implies that $u_n \to u_0$ in $C[0, 1]$, we see by Theorem 2.12 that (iv) is satisfied.

With the above Lemmas 9.4 and 9.5 we have all assumptions of Theorem 6.9 satisfied.

Theorem 9.12

*Assume that **A9, A10** are satisfied. Then problem (9.30) has at least one weak solution.*

Exercise 9.18 Using Theorem 6.5 examine the existence of a weak solution to (9.30) for $a < \pi$.

9.8 On Some Application of a Direct Method

Finally we remark on the variational solvability of (1.1) containing a nonlinear term in the special case when $a = 0$ and $N = 1$. Namely, basing on some exposition from [21], we consider the following Dirichlet Problem: find a function $u \in H_0^1 (0, 1)$ such that the following equation is satisfied:

$$\begin{cases} -\ddot{u}(t) + f(t, u(t)) = g(t), & \text{for a.e. } t \in (0, 1) \\ u(0) = u(1) = 0. \end{cases} \tag{9.32}$$

Here (apart from some further growth conditions):

A11 $\quad g \in L^2 (0, 1)$ *and* $f : [0, 1] \times \mathbb{R} \to \mathbb{R}$ *is an* L^2-*Carathéodory function.*

Observe that $F : [0, 1] \times \mathbb{R} \to \mathbb{R}$ given by

$$F (t, x) = \int_0^x f (t, s) \, ds \text{ for a.e. } t \in [0, 1] \text{ and all } x \in \mathbb{R}$$

is a Carathéodory function as well. We see that $\frac{d}{dx} F (t, x) = f (t, x)$ for a.e. $t \in [0, 1]$ and all $x \in \mathbb{R}$. Additionally we will assume that:

A12 $\quad$ *there exist* $a \in L^\infty (0, 1)$, $b, c \in L^1 (0, 1)$ *such that* $\|a\|_{L^\infty} < \pi^2$ *and that for a.e.* $t \in [0, 1]$ *and all* $x \in \mathbb{R}$ *it holds*

$$F(t, x) \geq \frac{1}{2} a (t) x^2 + b (t) x + c(t). \tag{9.33}$$

Remark 9.4

Assumption $\|a\|_{L^\infty} < \pi^2$ is connected with the Poincaré Inequality and is required in order to prove the coercivity of the corresponding Euler action functional

$$J : H_0^1 (0, 1) \to \mathbb{R}$$

given by

$$J(u) = \frac{1}{2} \int_0^1 |\dot{u}(t)|^2 \, dt + \int_0^1 F (t, u (t)) \, dt - \int_0^1 g (t) u (t) \, dt. \tag{9.34}$$

Observe that $J = J_1 + J_2 + J_3$, all considered on $H_0^1 (0, 1)$, where

$$J_1(u) = \frac{1}{2} \int_0^1 |\dot{u}(t)|^2 \, dt, \quad J_2(u) = \int_0^1 F (t, u (t)) \, dt, \quad J_3(u) = - \int_0^1 g (t) u (t) \, dt.$$

(continued)

Remark 9.4 (continued)
The following

$$F(t, x) = -\frac{1}{4}\pi^2 t x^2 + (\sin t) x$$

serves as an example of a function satisfying **A12** and the associated nonlinear term is

$$f(t, x) = -\frac{1}{2}\pi^2 t x + \sin t.$$

In order to apply the Direct Method, Theorem 2.22, we need to demonstrate for functional J the following properties:

- *sequential weak lower semicontinuity;*
- *coercivity;*
- *Gâteaux differentiability;*
- *strict convexity (if one wishes to obtain uniqueness).*

In the following sequence of lemmas we show that the above are satisfied. Note that in considering the existence of solutions to (9.32) we will look for weak solutions which are critical points (9.34). We say that $u \in H_0^1(0, 1)$ is a weak solution to (9.32) if

$$\int_0^1 \dot{u}(t)\dot{v}(t)\, dt + \int_0^1 f(t, u(t))v(t)\, dt = \int_0^1 g(t) v(t)\, dt \qquad (9.35)$$

for all $v \in H_0^1(0, 1)$. We have:

Lemma 9.6
*Assume that **A11** holds. Then functional J is differentiable in the sense of Gâteaux on $H_0^1(0, 1)$.*

Proof Since J_3 is linear and bounded, it is obviously C^1. From Example 2.5 it follows that J_1 is C^1 as well. Thus it remains to consider functional J_2. Fix $u \in H_0^1(0, 1)$. For a given $v \in H_0^1(0, 1)$ we investigate the existence of the following limit:

$$\lim_{\varepsilon \to 0} \int_0^1 \frac{F(t, u(t) + \varepsilon v(t)) - F(t, u(t))}{\varepsilon}\, dt.$$

Obviously

$$\lim_{\varepsilon \to 0} \frac{F\left(t, u\left(t\right) + \varepsilon v\left(t\right)\right) - F\left(t, u\left(t\right)\right)}{\varepsilon} = f\left(t, u\left(t\right)\right) v\left(t\right)$$

for a.e. $t \in [0, 1]$. Using the fact that both u, v are continuous (and so are bounded by some constant, say $d > 0$) and the Lagrange Mean Value Theorem we obtain that for a.e. $t \in [0, 1]$

$$\left| \frac{F\left(t, u\left(t\right) + \varepsilon v\left(t\right)\right) - F\left(t, u\left(t\right)\right)}{\varepsilon} \right| \leq \max_{s \in [-d, d]} \left| f\left(t, s\right) \right| d.$$

Hence we can apply the Lebesgue Dominated Convergence Theorem.

From formula (9.35) defining the weak solution and from Lemma 9.6 we obtain at once the following result connecting solutions to (9.32) with critical point to functional (9.34):

Lemma 9.7
Assume that A11 holds. Then $u \in H_0^1(0, 1)$ is a critical point to functional (9.34) if and only if it satisfies (9.35).

Lemma 9.8
Assume that A11 holds. Then functional J is sequentially weakly lower semicontinuous on $H_0^1(0, 1)$.

Proof We see that J_3 is linear and bounded, and therefore sequentially weakly continuous. Functional J_1 is convex and continuous which by Theorem 2.20 implies that it is sequentially weakly lower semicontinuous. From Example 2.11 we see that J_2 is sequentially weakly continuous.

Lemma 9.9
Under assumptions A11, A12 functional J is coercive over $H_0^1(0, 1)$.

Proof Put

$$a_1 = \|a\|_{L^\infty}, \ b_1 = \int_0^1 |b(t)| dt, \ c_1 = \int_0^1 |c(t)| dt.$$

Observe that it holds by the Sobolev and the Poincaré Inequality

$$J_2(u) \geq \frac{1}{2} \int_0^1 a(t)|u(t)|^2 dt + \int_0^1 b(t)u(t) dt + \int_0^1 c(t) dt$$
$$\geq -\frac{1}{2\pi^2} a_1 \|u\|_{H_0^1}^2 - b_1 \|u\|_{H_0^1} - c_1 \text{ for any } u \in H_0^1(0,1). \tag{9.36}$$

Summing up we have

$$J(u) \geq \frac{1}{2}\left(1 - \frac{1}{\pi^2} a_1\right) \|u\|_{H_0^1}^2 + \left(-b_1 - \int_0^1 |g(t)|\right) \|u\|_{H_0^1} - c_1$$

for any $u \in H_0^1(0,1)$. Since $a_1 < \pi^2$, we see that J is coercive over $H_0^1(0,1)$.

Using Lemmas 9.8, 9.6, and 9.9 we can apply Theorem 2.22 to reach the following result:

Theorem 9.13
Assume conditions A11, A12. Then problem (9.32) has at least one solution $u \in H_0^1(0,1) \cap H^2(0,1)$.

It remains to comment that the uniqueness is reached in case functional J has exactly one critical point. Hence we need one additional assumption:

A13   *for a.e. $t \in [0,1]$ function $x \mapsto f(t,x)$ is nondecreasing on $\mathbb{R}$.*

Theorem 9.14
Assume that A11, A13 hold. Then Dirichlet Problem (9.32) has exactly one solution $u \in H_0^1(0,1) \cap H^2(0,1)$.

Proof From **A13** it follows that for a.e. $t \in [0,1]$ function $x \mapsto F(t,x)$ is convex on $\mathbb{R}$ and therefore functional J_2 is convex as well. Since J_1 is strictly convex and J_3 convex, we see that now J is strictly convex and therefore its critical point, which exists by Theorem 9.13, is unique. The coercivity is obtained directly.

Exercise 9.19 Let $p > 2$. Assume that function $f : \mathbb{R} \to \mathbb{R}$ is continuous and nondecreasing. Let $F : \mathbb{R} \to \mathbb{R}$ be defined by

$$F(x) = \int_0^x f(s)\, ds.$$

Prove that the Dirichlet problem

$$\begin{cases} -\frac{d}{dt}\left(\left|\frac{d}{dt}u\left(t\right)\right|^{p-2}\frac{d}{dt}u\left(t\right)\right) + f\left(u\left(t\right)\right) = 0, \text{ for } a.e.\ t \in (0, 1) \\ u\left(0\right) = u\left(1\right) = 0 \end{cases}$$

has exactly one weak solution $u \in W_0^{1,p}\left(0, 1\right)$ which is a minimizer to the following action functional $J : W_0^{1,p}\left(0, 1\right) \to \mathbb{R}$ given by the formula

$$J\left(u\right) = \frac{1}{p}\int_0^1 \left|\dot{u}\left(t\right)\right|^p dt + \int_0^1 F\left(u\left(t\right)\right) dt.$$

References

1. R.A. Adams, *Sobolev Spaces* (Academic Press, London, 1975)
2. R.P. Agarwal, *Difference Equations and Inequalities: Theory, Methods and Applications* (Marcel Dekker, New York, 2000)
3. H.H. Bauschke, P.L. Combettes, *Convex Analysis and Monotone Operator Theory in Hilbert Spaces*. CMS Books in Mathematics/Ouvrages de Mathématiques de la SMC (Springer, Cham, 2017)
4. S. Boyd, E.K. Ryu, A primer on monotone operator methods (survey). Appl. Comput. Math. **15**(1), 3–43 (2016)
5. H. Brézis, *Functional Analysis, Sobolev Spaces and Partial Differential Equations.* (Springer, Berlin, 2010)
6. C. Canuto, A. Tabacco, *Mathematical Analysis I&II* (Springer, Berlin, 2008)
7. J. Chabrowski, *Variational Methods for Potential Operator Equations* (De Gruyter, Berlin, 1997)
8. R. Chiappinelli, D.E. Edmunds, Remarks on Surjectivity of Gradient operators. Mathematics **8**, 1538 (2020). https://doi.org/10.3390/math8091538
9. I. Cioranescu, *Geometry of Banach Spaces, Duality Mappings and Nonlinear Problems* (Kluwer Academic Publishers, Dordrecht, 1990)
10. Z. Denkowski, S. Migórski, N.S. Papageorgiou, *An Introduction to Nonlinear Analysis: Theory* (Kluwer Academic/Plenum Publishers, New York, 2003)
11. Z. Denkowski, S. Migórski, N.S. Papageorgiou, *An Introduction to Nonlinear Analysis: Applications* (Kluwer Academic/Plenum Publishers, New York, 2003)
12. G. Dinca, P. Jeblean, Some existence results for a class of nonlinear equations involving a duality mapping. Nonl. Anal. **46**, 347-363 (2001)
13. G. Dinca, P. Jebelean, J. Mawhin, Variational and topological methods for Dirichlet problems with p-Laplacian. Port. Math. (N.S.) **58**(3), 339–378 (2001)
14. P. Drábek, J. Milota, *Methods of Nonlinear Analysis. Applications to Differential Equations*, 2nd edn. Birkhäuser Advanced Texts. Basler Lehrbücher (Springer, Basel, 2013)
15. I. Ekeland and R. Temam, *Convex Analysis and Variational Problems* (North-Holland, Amsterdam, 1976)
16. D.G. Figueredo, *Lectures on the Ekeland Variational Principle with Applications and Detours.* Preliminary Lecture Notes (SISSA) (1988)
17. J. Franců, Monotone operators: a survey directed to applications to differential equations. Aplikace Matematiky **35**(4), 257–301 (1990)
18. S. Fučik, A. Kufner, Nonlinear differential equations, in *Studies in Applied Mechanics*, vol. 2 (Elsevier Scientific Publishing Company, Amsterdam, 1980)
19. H. Gajewski, K. Gröger, K. Zacharias, *Nichtlineare Operatorgleichungen und Operatordifferentialgleichungen* (Akademie, Berlin, 1974)
20. M. Galewski, On the application of monotonicity methods to the boundary value problems on the Sierpinski gasket. Numer. Funct. Anal. Optim. **40**(11), 1344–1354 (2019)

© The Author(s), under exclusive license to Springer Nature Switzerland AG 2021
M. Galewski, *Basic Monotonicity Methods with Some Applications*,
Compact Textbooks in Mathematics, https://doi.org/10.1007/978-3-030-75308-5

21. M. Galewski, *Wprowadzenie do metod wariacyjnych* (Wydawnictwo Politechniki Łódzkiej, Łódź, 2020). ISBN 978-83-66287-37-2
22. M. Galewski, On variational nonlinear equations with monotone operators. Adv. Nonlinear Anal. **10**, 289–300 (2021)
23. M. Galewski, J. Smejda, On variational methods for nonlinear difference equations. J. Comput. Appl. Math. **233**(11), 2985–2993 (2010)
24. K. Goebel, S. Reich, *Uniform Convexity, Hyperbolic Geometry, and Nonexpansive Mappings* (Marcel Dekker, New York, 1984)
25. M. Haase, *Functional Analysis: An Elementary Introduction*. Graduate Studies in Mathematics, vol. 156 (AMS, New York, 2014)
26. D. Idczak, A. Rogowski, On the Krasnosel'skij theorem—a short proof and some generalization. J. Aust. Math. Soc. **72**(3), 389–394 (2002)
27. A.D. Ioffe, V.M. Tikhomirov, *Theory of Extremal Problems* (in Russian). Series in Nonlinear Analysis and its Applications. Izdat., Nauka, Moscow (1974); see also: Theory of extremal problems. Translated from the Russian by Karol Makowski. Studies in Mathematics and its Applications, vol. 6 (North-Holland Publishing Co., Amsterdam, 1979), xii+460 pp
28. N. Iusem, D. Reem, S. Reich, Fixed points of Legendre-Fenchel type transforms. J. Convex Anal. **26**, 275–298 (2019)
29. J. Jahn, *Introduction to the Theory of Nonlinear Optimization*. 3rd edn. (Springer, Berlin, 2007)
30. R.I. Kačurovskiĭ, Monotone operators and convex functionals. Uspekhi Mat. Nauk **15**, 213–215 (1960)
31. R.I. Kačurovskiĭ, Nonlinear monotone operators in Banach spaces. Uspekhi Mat. Nauk **23**, 121–168 (1968); English translation: Russian Math. Surveys **23**, 117–165 (1968)
32. G. Kassay, V.D. Rădulescu, *Equilibrium Problems and Applications* (Academic Press, Oxford, 2019)
33. W.G. Kelley, A.C. Peterson, *Difference Equations: An Introduction with Applications*, 2nd edn. (Harcourt/Academic Press, San Diego, 2001)
34. A. Kristály, V.D. Rădulescu, C. Varga, *Variational Principles in Mathematical Physics, Geometry, and Economics: Qualitative Analysis of Nonlinear Equations and Unilateral Problems*. Encyclopedia of Mathematics and its Applications, vol. 136 (Cambridge University, Cambridge, 2010)
35. J. Mahwin, *Problemes de Dirichlet Variationnels Non Linéaires*. Séminaire de Mathématiques Supérieures, Montreal, vol. 104 (1987)
36. A. Matei, M. Sofonea, *Variational Inequalities with Applications: A Study of Antiplane Frictional Contact Problems*. Advances in Mechanics and Mathematics, vol. 18 (Springer, New York, 2009)
37. S. Migórski, M. Sofonea, Variational–Hemivariational inequalities with applications, in *Chapman & Hall/CRC Monographs and Research Notes in Mathematics, Boca Raton, FL* (2018)
38. G.J. Minty, Monotone (nonlinear) operators in Hilbert space. Duke Math. J. **29**, 341–346 (1962)
39. G.J. Minty, On a "monotonicity" method for the solution of nonlinear equations in Banach spaces. Proc. Natl. Acad. Sci. USA **50**, 1038–1041 (1963)
40. G. Molica Bisci, D.D. Repovš, On some variational algebraic problems. Adv. Nonlinear Anal. **2**(2), 127–146 (2013)
41. G. Molica Bisci, V.D. Rădulescu, R. Servadei, Variational methods for nonlocal fractional problems, in *Encyclopedia of Mathematics and its Applications*, vol. 162 (Cambridge University, Cambridge, 2016)
42. D. Motreanu, V.D. Rădulescu, *Variational and Nonvariational Methods in Nonlinear Analysis and Boundary Value Problems, Nonconvex Optimization and Its Applications* (Springer, Berlin, 2003)
43. J. Nečas, Introduction to the theory of nonlinear elliptic equations, in *Teubner-Texte zur Math*, vol. 52 (Teubner, Leipzig, 1983)
44. N.S. Papageorgiou, V.D. Rădulescu, D.D. Repovš, Nonlinear analysis—theory and methods, in *Springer Monographs in Mathematics* (Springer, Cham, 2019)

45. R.R. Phelps, *Convex Functions, Monotone Operators and Differentiability*, 2nd edn. Lecture Notes in Mathematics, vol. 1364 (Springer, Berlin, 1993)
46. T.L. Rădulescu, V.D. Rădulescu, T. Andreescu, *Problems in Real Analysis: Advanced Calculus on the Real Axis* (Springer, New York, 2008)
47. V.D. Rădulescu, D.D. Repovš, *Partial Differential Equations with Variable Exponents: Variational Methods and Qualitative Analysis* (CRC Press/Taylor and Francis Group, Boca Raton, 2015)
48. D. Reem, S. Reich, Fixed points of polarity type operators. J. Math. Anal. Appl. **467**, 1208–1232 (2018)
49. S. Reich, Book review: geometry of banach spaces, duality mappings and nonlinear problems. Bull. Am. Math. Soc. (N.S.) **26**, 367–370 (1992)
50. M. Renardy, R.C. Rogers, *An Introduction to Partial Differential Equations*, 2nd edn. Texts in Applied Mathematics, vol. 13 (Springer, Heidelberg, 2004)
51. T. Roubíček, *Nonlinear Partial Differential Equations with Applications.* (International Series of Numerical Mathematics, vol. 153) (Birkhäuser, Basel, 2013)
52. W. Rudin, *Principles of Mathematical Analysis*, 2nd edn. (McGraw-Hill Book Co., New York, 1964)
53. W. Rudin, Functional analysis, in *McGraw-Hill Series in Higher Mathematics* (McGraw-Hill Book Co., New York, 1973)
54. R.E. Showalter, *Monotone Operators in Banach Space and Nonlinear Partial Differential Equations* (American Mathematical Society, Providence, RI, 1997)
55. T. Tao, An introduction to measure theory, in *Graduate Studies in Mathematics*, vol. 126 (2011)
56. J.L. Troutman, Variational calculus and optimal control, in *Optimization with Elementary Convexity, Undergraduate Texts in Mathematics* (Springer, New York, 1996)
57. E.H. Zarantonello, Solving functional equations by contractive averaging, in *Mathematical Research Center Technical Summary Report no. 160* (University of Wisconsin, Madison, 1960)
58. E. Zeidler, *Nonlinear Functional Analysis and Its Applications II/B—Nonlinear Monotone Operators* (Springer, New York, 1990)

Index

A
Adjoint operator, 64
Algebraic equation, 12
Arzela–Ascoli Theorem, 31

B
Banach Contraction Principle, 107
Banach–Steinhaus Theorem, 62
Bounded bilinear form, 143
Brouwer Fixed Point Theorem, 9
Browder–Minty Theorem, 112

C
Carathéodory function, 37
Chain Rule, 42
Clarkson Inequality, 33
Condition
    (M), 116
    (S), 75
    $(S)_+$, 76
    $(S)_0$, 76
    $(S)_2$, 76
Continuous
    demicontinuous, 66
    hemicontinuous, 66
    Lipschitz, 66
    radially, 66
    strongly, 66
    uniformly, 66
    weakly, 66
Convexity criteria, 44

D
Direct method of the calculus of variation, 51
Dirichlet Problem, 1
    for the generalized p–laplacian, 163
    for the laplacian, 156

Duality mapping, 127
Duality mapping relative to a function, 134
du Bois-Reymond Lemma, 35

E
Effective domain, 81
Ekeland Variational Principle, 152
Epigraph, 44

F
Fenchel-Moreau Theorem, 89
Fenchel-Young dual, 85
Fenchel-Young Inequality, 86
Fermat Rule, 40
Finite Dimensional Existence Theorem, 10
Fréchet derivative, 40
Fubini Theorem, 27

G
Galerkin method, 139
Galerkin solution, 138
Gâteaux derivative, 40
Gâteaux variation, 43
Generalized Browder–Minty Theorem, 117
Generalized Krasnosel'skii Theorem, 38

H
Hahn-Banach Theorem, 125
Hölder Inequality, 23
Hyperplane
    closed, 84
    supporting , 85

I
Invertibility of a monotone operator, 114
Iteration method, 109

K
Krasnosel'skii Theorem, 39

L
Lax-Milgram Theorem
   -nonlinear , 145
Lebesgue Dominated Convergence Theorem,
   37
Lemma
   du Bois-Reymond Lemma, 35
   fundamental in operator theory, 78
Leray–Lions Theorem, 121

M
Mean Value Theorem, 46
Minty Lemma, 77
   finite dimensional , 7
Monotone map, 6

N
Nemytskii operator, 37
Normalized duality mapping, 126

O
Operator
   bounded, 63
   coercive, 69
   d-monotone, 56
   locally bounded, 62
   monotone, 55
   potential, 91
   pseudomonotone, 118
   Riesz, 17
   strictly monotone, 55
   strongly monotone , 56
   uniformly monotone , 55
   weakly coercive, 69

P
Poincaré Inequality, 31
Pseudomonotone operator, 118

R
Riesz Representation in L^p, 22
Riesz Representation Theorem, 17
Ritz solution, 140

S
Separation Theorem, 83
Sequentially weakly closed, 17
Sequential weak lower semicontinuity, 47
Sobolev Inequality, 31
Strictly convex space, 21
Strongly Monotone Principle
   -general version, 118
Strongly Monotone Principle in a Banach
     space, 133
Sufficient conditions for monotonicity, 60
Sufficient convexity condition, 46

T
Theorem
   convexity and monotonicity, 59
   existence result for a pseudomonotone
     operator, 120
   sufficient condition for sequential weak
     lower semicontinuoity, 48

U
Uniformly convex space, 21

W
Weak convergence, 16
Weak derivative, 24
Weak sequential compactness of a ball, 18
Weierstrass Theorem, 47

Printed in the United States
by Baker & Taylor Publisher Services